天津解放桥（昔日万国桥），穿过这座老桥即可来到百年闻名的老街解放路
The Tianjin Jiefang Bridge with a century of history, leads across to Jiefang Road.

经历了半个多世纪的风霜雨雪，风采依旧
After a century of wind, rain and snow, Jiefang Road still keeps its grace and dignity

银行门前的罗马柱象征着庄严、稳定,将时间凝固成岁月
The Roman style columns in front of the bank are the symbols of dignity and stability

TIANJIN OLD BANKS

TIANJIN old BANKS

天津老银行

目 录
Contents

008
序
PREFACE

015
盐 业 银 行
SALT INDUSTRIAL BANK

035
汇 丰 银 行 天 津 分 行
HONGKONG & SHANGHAI BANKING CORPORATION, TIANJIN BRANCH

047
浙 江 兴 业 银 行 天 津 分 行
NATIONAL COMERCIAL BANK, TIANJIN BRANCH

065
横 滨 正 金 银 行 天 津 分 行
YOKOHAMA SPECIE BANK, TIANJIN BRANCH

081
中 法 工 商 银 行 天 津 分 行
INDUSTRIAL AND COMMERCIAL BANK OF CHINA-FRANCE, TIANJIN BRANCH

097
中 央 银 行 天 津 分 行
CENTRAL BANK, TIANJIN BRANCH

111
华 俄 道 胜 银 行 天 津 分 行
RUSSO-CHINESE BANK, TIANJIN BRANCH

125
金 城 银 行
KINCHEN BANK

139
麦 加 利 银 行 天 津 分 行
THE CHARTERED BANK OF INDIA, AUSTRALIA AND CHINA, TIANJIN BRANCH

153
四行储蓄会天津分会
THE NORTHERN FOUR SAVING BANKING SOCIETY, TIANJIN BRANCH

165
花旗银行天津分行
INTERNATIONAL BANKING CORPORATION, TIANJIN BRANCH

175
东莱银行天津分行
TUNG LAI BANK, TIANJIN BRANCH

193
华比银行天津分行
CHINA-BELGIUM BANK, TIANJIN BRANCH

207
大陆银行
MAINLAND BANK

221
中南银行天津分行
CHINA & SOUTH SEA BANK, TIANJIN BRANCH

233
朝鲜银行天津分行
KOREAN BANK, TIANJIN BRANCH

243
东方汇理银行天津分行
INDO-CHINA AGENCY BANK, TIANJIN BRANCH

259
新华信托储蓄银行天津分行
XIN HUA TRUST & SAVINGS BANK, TIANJIN BRANCH

272
附录
APPENDIX

283
老银行——天津近代经济的核心
OLD BANKS — CENTRAL TO TIANJIN'S ECONOMIC GROWTH

292
跋
POSTSCRIPT

296
后记
POSTSCRIPT

天津老银行

在近代中国历史上，天津就是北方的经济中心，重要的商贸中心、金融中心和口岸城市。18世纪初，由于埠际商业的发展，天津出现了办理汇兑业务的钱庄，商贸业与金融业的相互促进，使天津逐渐发展成为船桅林立、客商云集的繁华口岸。第二次鸦片战争后，天津被迫开放为通商口岸，这一特殊历史时期，各国的租界和洋楼密布天津。资料显示，在上个世纪30年代，天津的对外贸易量相当于全中国的四分之一，这个数量是很大的。当时，天津的金融业已经比较发达，现在的天津解放路一带银行林立、盛极一时，被称为中国"华尔街"。1.8 km长的路段两侧曾聚集了近30家银行，中国第一个证券交易所也设在这里。据史料载，在20世纪40年代，天津设有7家银行总行，33家银行分行，269家银号，17家外资银行，50多家保险公司和228家外商保险机构，还有一家证券交易所。当时，在天津的华商银行资本总额占全国的12.7%，外国银行资本总额占全国外国银行资本总额的16%，银行营业额占全国的18%。

新中国成立以后，我市经济和金融业发展的外部条件发生了很大的变化。改革开放以来，特别是近年来，天津市委、市政府高度重视金融工作，制定并实施了一系列有利于金融业发展的政策和办法，为加快金融改革与发展创造了良好的政策环境和社会环境，天津金融业进入发展最好、最快的时期之一，为全市经济和社会发展作出了重要贡献。全市金融业务总量迅速增长，金融机构体系基本形成，金融生态环境不断改善，金融服务功能全面增强，金融改革创新迈出坚实步伐，金融业对外开放不断扩大，金融资产质量明显提高，金融业秩序稳定，呈现出强大发展活力和强劲的发展势头。

2006年5月，国务院下发《关于推进天津滨海新区开发开放有关问题的意见》，批准天津滨海新区为全国综合配套改革试验区，并把金融改革创新列为综合配套改革之首，允许在金融企业、金融业务、金融市场和金融开放等方面先行先试，为全国金融业的发展提供经验和借鉴。根据天津的发展实际，市委、市政府认真研究后提出，我市金融业发展目标是，到2010年建成与北方经济中心相适应的现代金融服务体系，办好全国金融改革创新基地。当前，天津正在围绕拓宽直接融资渠道、开展金融企业综合经营、组建创新型金融机构和深化外汇管理改革等重点大力推进金融改革创新。新形势下，天津金融业发展面临着艰巨任务，担负着重大使命。我们要深入贯彻科学发展观，认真贯彻落实党的十七大精神，不断开创我市金融改革发展的新局面，为把天津建成我国北方经济中心、国际港口城市和生态城市作出新贡献。

如今，风格各异的天津老银行建筑，历经近百年沧桑，屹立在解放路两侧。它们记录着天津金融业的辉煌昨天，是历史留给天津的宝贵财富和独特资源，也是天津作为历史文化名城的重要载体。

《天津老银行》这部书，通过大量实景照片和丰富史料，对近代天津金融建筑进行了全面梳理和翔实介绍，展现了天津老银行的丰富历史内涵和厚重文化底蕴。它有利于我们合理开发、认真保护、有效利用历史风貌建筑，深入挖掘天津老银行的丰厚文化内涵，无论对于天津近代经济和金融研究，对于促进天津现代金融业和旅游业发展，还是更好地开展对外经济合作、文化交流，都具有积极深远的意义。

是为序。

2007.12

Tianjin Old Banks

In modern Chinese history, Tianjin, a port city of northern China, has been an important economic and trade center.

Thanks to the expansion of inter-port business, banking houses offering remittance services, appeared in Tianjin in the early 18th century. The co-development of commerce and finance turned Tianjin into a busy and flourishing port crowded with ships and merchants from all over China. After the Second Opium War, Tianjin was forced to open as an international trading port. During that special historical period, foreign concessions and Western-style buildings were seen in many areas of the city. Statistics show that in the 1930s, the trade volume of Tianjin accounted for one-fourth of the country's total, a very significant amount. By then, Tianjin was already relatively developed in its financial industry. The area of Jiefang Road was full of banks. It was such a financially prosperous area that it was dubbed "China's Wall Street". On both sides of a road, which was 1.8 km long, stood about 30 banks. The first foreign exchange of China was also located here. It was recorded that in the 1940s seven banks had their head offices, and 33 banks had their branches in Tianjin. Tianjin boasted 267 money houses, over 50 insurance companies, 228 foreign insurance institutions and a securities exchange. Of all the capital in Chinese banks, 12% was held by the branches in Tianjin; while foreign banks in the city held 16% of the total holdings of all foreign banks in China.

After the founding of the People's Republic of China, Tianjin experienced great changes in the external conditions of its economic and financial development. Since the reform and open policies were adopted, particularly in recent years, Tianjin Municipal Party Committee and Tianjin Municipal Government, have put great emphasis on the finance industry and have drawn up and implemented a series of favorable policies and systems. Sound policies have been made and a healthy environment has been created to facilitate financial reforms and expansion. The financial industry is now in a peak period and has made significant contribution to the city's economic and social development. The total volume of its financial business has risen rapidly in recent years. A complete spectrum of financial institutions has by and large been established. With constant improvement in the city's environment and infrastructure, the overall functionality of its financial services has been strengthened. With the financial industry being more open to the outside world, Tianjin has seen improvement in the quality of its financial assets. Having put its financial industry in to good order, the city is exhibiting great vitality and a created a strong momentum for development.

In May 2006, the Chinese State Council issued a document, "Comments on Relevant Issues Concerning Development and Opening up in Tianjin Binhai New

Area", approving the establishment of a special economic reform zone, which will assist the overall development of China. Giving top priority to reforms and innovation in the financial sector, the Chinese State Council has granted Tianjin permission to take the lead in undertaking experiments in opening up financial enterprises, financial services, financial market and the financial sector, so that Tianjin might provide valuable experience to other parts of the country. The Municipal Party Committee and Municipal Government have set a goal of setting-up a system of modern financial services compatible with its status as the economic center of Northern China and build itself into a successful national base for financial reforms and innovation by the year 2010. At present, Tianjin is focusing on such priorities as expansion of direct financing channels, promotion of coordinated management of financial enterprises, establishment of innovative financial institutions and intensification of reforms in foreign exchange management. Being entrusted with the development of its financial industry Tianjin is facing a huge challenge ahead. In the spirit of the 17th National People's Congress, the Municipality will make a sustained effort to create new opportunities for financial reforms in a bid to make a renewed contribution to building Tianjin into a northern economic center, an international port and an ecological city.

Today, the buildings of the old banks in Tianjin, with their diverse architectural style, still stand along Jiefang Road after nearly a hundred years of change. They are the recorded history of the glory of the city's financial industry.

The book, "Tianjin Old Banks", presents a collection of pictures showing a realistic view of the old banks and providing a rich historical documentation of the cities banks. They are an accurate account of the architecture of the financial institutions that form part of modern-day Tianjin. They serve to reveal the historical richness and cultural value of the buildings. The book gives us the possibility to understand these historical buildings, an important step if we are to act to protect and utilize them in a manner which is in harmony with their cultural significance. Further, it is hoped the book will have a positive impact, inspiring others to study the economic and financial history of the city and encourage future development of the city's financial industry and tourism, enhancing economic cooperation and cultural exchange between Tianjin and other countries.

Dai Xianglong

2007.12

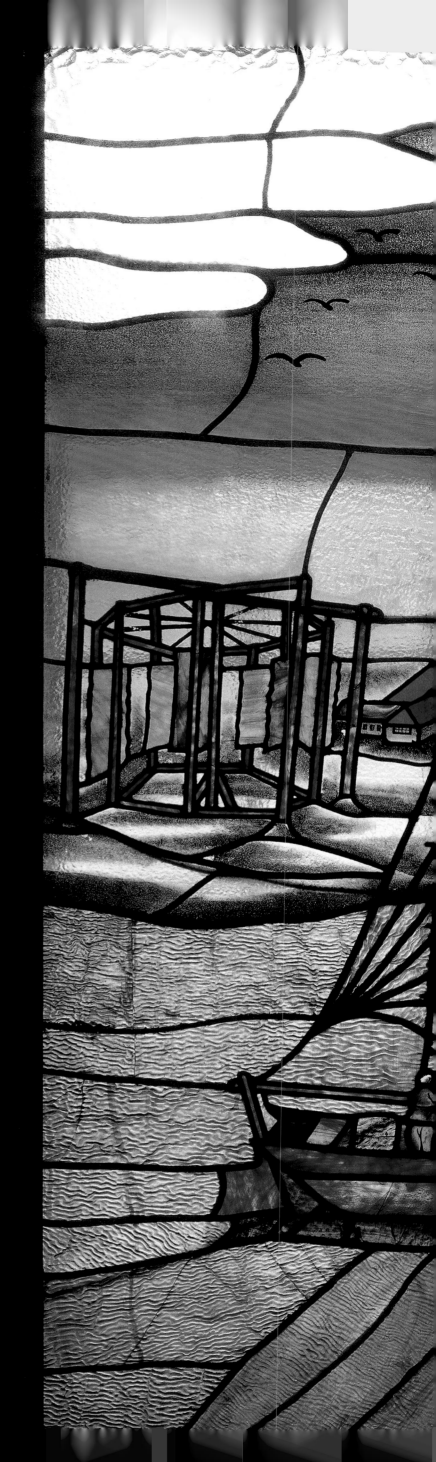

门镶有商船运盐景象的比利时彩色玻璃窗
ell with Belgian stained glass depicting scenes of merchant ships and the salt trade

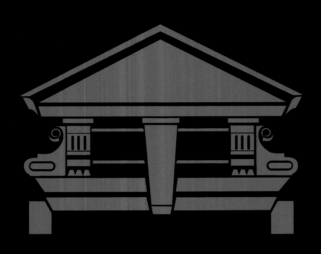

昔日盐业银行（今日的中国工商银行天津分行）
Formerly Salt Industrial Bank (Now Commercial and Industrial Bank of China, Tianjin Branch)

古典风格的窗户
Classical style windows.

盐业银行
SALT INDUSTRIAL BANK

盐业银行是北洋政府本着以盐款为财政收入大宗，为维持盐业、调剂金融，由国务卿徐世昌、财政部参政张镇芳筹办。1915年3月26日成立，总行设于北京。当年5月29日盐业银行天津分行成立。该行创立时为官商合办，资本额500万元，其中官股200万元，开业时实收64万元，其中官股只有10万元。1917年该行改为商办。1923年改定资本额为1 000万元，实收700万元，成为全国商业银行之冠。1928年8月盐业银行总行迁往天津，行址在法租界8号路（今赤峰道，原天津分行行址）。1934年6月盐业银行总行又迁往上海。1952年12月该行参加全行业公私合营，与其他银行、钱庄一起组成统一的公私合营银行。

盐业银行在津经收天津地区全部盐税收入，办理存放款、汇总、储蓄等商业银行业务。

盐业银行旧址位于天津市和平区赤峰道12号，由华信工程司沈理源设计，建于1926年。占地面积3 174m²，建筑面积6 244m²，建筑形式为罗马古典复兴形式，建筑以半地下室为台基。一、二层用6根罗马混合式巨柱式空柱廊统一起来。罗马混合式柱头上的圆形涡卷，吸取中国方形回纹来代替，将东西方文化结合得十分巧妙，具有独创性。入口柱廊部有山花，入口大门两旁各有一根壁柱及一根带有回纹的混合式巨柱，同样的立面处理运用于沿街立面另一拐角处。形成一个镂空柱廊。正面三米高的大铜门，金光闪闪。从转角入口上八步台阶进入营业大厅，大厅的柜台沿大厅三面设置，另一面为正八角形的会客室。在中间形成长八角形的空间。天棚用黄金等材料制成"蓝天飞凤满天星"图案。天棚中间为蓝天，四角有凤凰的浮雕像。地面均用意大利产大理石铺地。地面设有用白大理石雕刻的座椅。柜台内天棚用小八角形藻井，镶嵌彩色玻璃。营业大厅内沿两侧柜台处各靠有四根类似混合式柱子，白色的柱头，深色大理石的圆柱子。大厅两端各有四根方柱子，使整个营业大厅雄伟庄重，富丽堂皇。营业大厅西北为楼梯间，其窗户镶有盐滩晒盐及帆船运盐景象的彩色玻璃，富有天津地方特色。二楼有宴会厅及会议厅，办公室等。装饰华丽，三楼为职员宿舍。在近代曾因藏有16只国宝金编钟而闻名。该建筑经国务院批准为国家级文物保护单位。

上：入口两侧华丽的大理石柱
Above: The magnificent marble entrance columns

下：厅内的吊灯
Below: The chandelier in the Grand Hall

各屋悬挂的时钟已近百年，每当调整时间，只需调整一个时钟其他各屋都会随之改变

The clocks in the room are almost a hundred years old. If one clock is adjusted all the other clocks in the other rooms will change as well

During the Northern Warlords period, salt tax formed the bulk of the public revenue, state secretary Xu Shichang and Ministry of Finance supervisor Zhang Zhenfang established the Salt Industrial Bank to maintain the salt industry and allocate funds. Founded on March 26th 1915 with it's headquarter in Beijing. May 29th of that year saw the founding of the Tianjin Branch. Initially the bank was a government-private joint bank with the capital amount of 5 million Yuan, of which the governmental share was 2 million Yuan. When it was open, the paid-up capital was 640,000 Yuan with a government share of only 100,000 Yuan. In 1917, the bank was privatized. In 1923, the

楼梯间内镶有晒盐景象的彩色玻璃窗

Stairwell with stained glass scene of the sun drying salt

装饰华丽的休息厅
The magnificent common room

capital amount was changed into 10 million Yuan with paid-up capital of 7 million Yuan, making it the highest ranking bank among the Chinese commercial banks. In August 1928, its headquarters moved to Tianjin, to Rue Pasteur in the French Settlement (now Chifeng Road). In June 1934, the headquarters moved back again to Shanghai. In December 1952, the bank was nationalized by the communist government.

The office of the Salt Industrial Bank was designed by Shen Liyuan of Hua Xin Engineering Company. Located at 12 Chifeng Road and constructed in 1926, the building's plan is rectangle-shaped, and it is a 3-story mixed structure with basement. Its covering area is 3,174 m² with a floor space of 6,244 m². The building is in the form of Roman Classical Revival. With semi basement as the base, the first and the second floor are unified by 6 huge Roman colonnades. The Roman column capital was replaced by Chinese style square-shaped winding pattern, masterly combining eastern and western cultures. Huge columns with winding pattern in the capital stand at each side of the 3 meters high copper entrance door. The corner entrance leads to the business hall. The cashier counters are set up along the three sides of the hall, at the fourth side is an octagon-shaped reception room. Original the ceiling was painted with the scene "flying phoenix in the blue sky full of stars" which included gold details. In the center of the ceiling, there is the pattern of blue sky, and at the 4 corners relief sculptures of the phoenix. Italian marble covers the floor, where sit delicately-sculptured white marble Chairs. Inside the counter, a small octagon-shaped Chinese ceiling is used on the top with stained glass inlaid. There are 4 square columns at each side of the hall making the whole business hall grand and magnificent. To the northwest of the business hall is the staircase, where stained glass windows depict "sunning the salt on the salt flat" and "sailing boat carrying salt" with full Tianjin characteristics. On the second floor is the Banquet Hall, Conference Hall and other offices, all with fabulous decoration. The employees' dormitory used to be on the third floor. In more recent history, it is famous for protecting and hiding the national treasure—16 golden chime bells. The state council lists this fine example of architecture as a national cultural relic.

营业大厅侧面镶嵌着彩色玻璃的小八角藻井
The caisson ceiling in the business hall, decorated by stained glass

上：厅内造型各异的大理石柱
Above: The marble columns in the hall are carved in various styles

下：会客室内八角形的桌子与八角形的屋顶上下呼应，混为一体
Below: The octagonal table matches the octagonal ceiling inside the Meeting Room

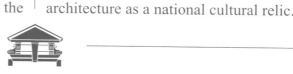

大理石桌子
Marble table

柜台内通往地下室的走廊
Behind the counter a corridor leading to the basement

八角形藻井布满天棚
ss hall with octagonal caisson ceiling

精致的大理石雕柱头
Exquisite marble column capital

八角形的经理室，由此可一览营业大厅全貌
The octagonal manager's office, with view of the whole business hall

楼梯间和通往办公室的走廊
The stairwell and corridor leading to the offices

会客室
Reception room

会议厅一角
Meeting hall

走廊一角
Corridor

华丽的会议大厅
The magnificent meeting hall

HONGKONG & SHANG-
HAI BANKING CORPORA-
TION, TIANJIN BRANCH

汇丰银行

天津分行

昔日的汇丰银行天津分行（今日的中国银行天津分行）
The former HSBC Tianjin Branch (Now Bank of China, Tianjin Branch)

斜阳下的艾奥尼克式柱廊
The Ionic colonnade at sunset

汇丰银行
天津分行
HONGKONG & SHANGHAI BANKING CORPORATION, TIANJIN BRANCH

英国汇丰银行创办于1864年，总行设于香港。清光绪七年（1881年）汇丰银行天津分行在英租界宝士徒道（今营口道）开业，是最早在天津设立的外国银行。1925年迁入中街新楼（今解放北路86号）营业，直至1941年12月太平洋战争爆发，被日军接管停业。1945年抗战胜利后，在原址恢复营业，至1954年撤离。

发行钞票是汇丰银行天津分行的主要业务之一。汇丰银行天津分行先后发行加印汇丰银行天津地名纸币有1墨西哥银元、25墨西哥银元、50墨西哥银元、5元和10元共计7个版别。据1933年统计，汇丰银行天津分行发行钞票达47.5万元。

汇丰银行在天津开设分行后，借助地理之便，与当时的清政府关系保持密切。从1880年到1927年，汇丰银行天津分行一共借给清政府款项多达48笔，借出白银3.38亿两，取得了盐业与海关两大税源的总管权。汇丰银行天津分行掌管着天津的国际汇兑业务，把持指挥外商国际汇兑银行公会和外汇经纪人公会，天津每天的外汇牌价都是以其挂牌为准。1934年天津共有外商银行17家，资产总额436 129 649元，汇丰银行天津分行占18%。

门厅一角
Entrance hall

位于解放北路86号的大楼遗址为1925年重建的汇丰银行天津分行大楼。该建筑于1924年由阿特金森(Athinson)及达拉斯(Dallas)的同和工程司设计，设计者为该工程司的苏格兰建筑师伯内特(B.C.G.Burnett)，1925年建成，占地4 400m²，建筑面积5 539m²。该建筑设计风格为希腊古典复兴式，东立面入口为四根爱奥尼克巨柱式门廊，立面三段划分明确，有台基、柱子和檐部，檐部与柱高比约为1:4。上部为三角形山花图案。两侧旁门有两根塔什干圆柱承托的短檐部的门廊。各入口处均为花饰铜大门，南立面是8根爱奥尼克巨柱式门廊。中央突出的4根爱奥尼克巨柱式柱廊，形式与东立面柱廊相同。中央柱廊两侧均有略向后退的两根爱奥尼克巨柱式的空柱廊，从而使中央柱廊更为突出。首层营业厅面积为670m²。对称布置天然大理石圆柱。营业大厅屋顶为券柱式结构，中央采用井字梁，用双层玻璃顶棚，在井字梁天花内镶嵌着钢丝网彩色玻璃，使大厅既采光明亮又美观。柜台外采用大理石地面，柜台内为软木地板，均由美国肯尼迪公司安装。大厅四周为办公室、保险库、账库、卫生间等，二、三层为办公室、会客室、宿舍等。地下室建有金库及保险库。由于解放北路与大同道为斜交，故入口门厅布置成椭圆形，巧妙地使入口门厅与营业大厅的轴线不在一条直线上，又不为人们所察觉。该建筑系天津市文物保护单位，特殊保护等级历史风貌建筑。现为中国银行天津分行。

上：藻井檐边的狮头造型浮雕
Above: Lion head relief under the caisson ceiling

下：银行大楼外景
Below: Exterior view of the Bank

汇丰银行南门外景
View of the south entrance of HSBC

British Hong Kong and Shanghai Banking Corporation, headquartered in Hong Kong, was founded in 1864. In 1881, the 7th year of Kuang Tsu of Qing Dynasty, Tianjin Branch of Hong Kong and Shanghai Banking Corporation opened on Bristow Road in the British Settlement (now Yingkou Road). It was the first foreign bank in Tianjin. In 1925, it moved to the new building on Victoria Road (now No.82 Jiefang North Road). Having been taken over by the Japanese army, it terminated business after the outbreak of Pacific War in December, 1941. After the victory in the Anti-Japanese War in 1945, it resumed its business at the original site and but closed in 1954.

Issuing banknotes was the mainstream businesses of the Tianjin Branch of the Hong Kong and Shanghai Banking Corporation. It successively issued 7 versions of paper money imprinted with the local name Tianjin, including 1 Mexico silver dollar,

Stained glass window in the grand hall

Exquisite wooden staircase

25 Mexico silver dollars, 50 Mexico silver dollars, 5 Yuan and 10 Yuan. According to the statistics of 1933, the banknotes issuing amount of Tianjin Branch of the Hong Kong and Shanghai Banking Corporation reached 475,000 Yuan.

After the founding of Tianjin Branch, it maintained, by virtue of its geographic proximity, close relationship with the Qing government. From 1880 to 1927, it had made 48 loans to the Qing government totaling 338 million liang of silver, and acquired a monopoly of tax sources from the salt industry and customs. Tianjin Branch controlled international exchange business in Tianjin, commanded foreign exchange bankers' association and foreign exchange brokers' association. Its quote price was taken as standard for Tianjin daily foreign exchange quote price. Up to 1934, there were totally 17 foreign banks in Tianjin, whose total assets amounted to 436,129,649 Yuan, 18% of which was attributed to the Tianjin Branch of the Hong Kong and Shanghai Banking Corporation.

Sited on No.84 Jiefang North Road, it was established in 1925 and the building was designed by B.C.C. Burnet of British Atkinson & Dallas Engineering Company, with an area of 4,400 m² and floor space of 5,539 m². It is of classical Greek form. At the entrance of the eastern elevation, there is the porch of 4 huge Ionic columns with classical proportions. On the top of east entrance, there are patterns of triangle-shaped pediments. The gate of each entrance is made of copper with flower ornaments. At the south elevation, 8 Ionic columns form a colonnade. 4 huge columns form the central colonnade, flanked at the two sides by 2 more columns that are set back a little. The effect makes the central colonnade more outstanding. The first floor is the business hall of 670 m², supported by 16 marble columns. The top of the hall is of arch and column structure with beams at the center and double layered glass ceiling. In between the beams is inlaid with stained glass, which makes the hall both bright and elegant. The main floor was of marble, with a cork floor behind the counters. The entire decor originated from the American Kennedy Corporation. Leading off the hall are offices, vaults, accounting rooms, washroom, etc. The exchequer and vaults were built in the basement. Since Jiefang North Road is intersected with Datong Road at an angle, the entrance hall is an elliptical shape meaning the entrance hall and the business hall are not square, a fact which is not obvious when viewed inside. The building is classed as a Tianjin cultural relic. It is currently occupied by Bank of China, Tianjin Branch.

Office fireplace

营业大厅内镶嵌着马赛克的地面
Mosaic floor in the business hall

气势恢弘的营业大厅
The magnificent business hall

大厅四周的带有短檐部的彩色大理石的托斯卡纳柱子细部
Tuscan columns around the grand hall

NATIONAL COMMERCIAL BANK, TIANJIN BRANCH

浙江兴业银行

天津分行

昔日的浙江兴业银行天津分行

浙江兴业银行
天津分行
NATIONAL COMMERCIAL BANK, TIANJIN BRANCH

浙江兴业银行在天津发行的纸币
Banknotes issued by National Commercial Bank, Tianjin Branch

1905年5月27日浙江兴业银行由浙江铁路公司创建成立，总行设在杭州。股金总额定为100万元，先收四分之一。同年设汉口、上海分行。1915年浙江铁路公司收归国有，铁路公司的股份转由私人商业资本承受，浙江兴业银行进行改组，收足资本金100万元，上海分行改为总行。同年10月24日浙江兴业银行天津支行成立，行址在宫北大街。1921天津支行改为天津分行。1925年浙江兴业银行天津分行行址迁到法租界21号路（今和平路、滨江道口）。1952年12月该行参加全行业公私合营，与其他银行、钱庄一起组成统一的公私合营银行。

浙江兴业银行享有兑换券发行权（1935年废止）。1923年浙江兴业银行天津分行发行加印浙江兴业银行天津地名券1元、5元、10元共计3个版别。

浙江兴业银行与浙江实业银行、上海商业储蓄银行和新华信托储蓄银行并称为"南四行"，该行除办理一般商业银行业务外，兼办代理、租赁、仓库事宜。浙江兴业银行天津分行主要业务为存款、放款和汇兑，并兼营各种货币买卖及仓库业。

浙江兴业银行天津分行坐落在和平路319号，该建筑由华信工程司沈理源设计，1922年建成。为2层混合结构带半地下室，占地面积3 133m²，建筑面积2 034m²。建筑位于和平路与滨江道转角，建筑形式上古典主义建筑三段论很明确。底层作为台基、二、三层为柱身，四层为檐部及女儿墙，但入口上部二、三层为跨弧形的类似爱奥尼克双圆柱空廊，檐壁做成镶板，并用双牛腿支撑檐口，双圆柱墩处有瓶饰栏杆，入口底层门廊用塔司干双柱支撑。由于入口在马路转角处，所以用很深的门廊把台阶放置门廊内，以便与马路的行人有缓冲的余地。底层墙面用花岗岩块石，用深缝砌筑。底层窗户均用半圆拱窗，拱顶石用狮子头作雕饰，窗外有精美的铁花装饰，营业大厅内呈圆形，用14根深色大理石圆柱环绕，上设环形梁，梁上雕有中国古钱币图案，下设有营业用柜台，柜台用精美大理石雕刻，并用精美的狮子头雕饰支撑。大厅顶部用半球形钢网架支撑，白色磨花玻璃镶嵌，使大厅内显得十分明快，大厅两侧有经理室、会客室、文书、会计室等，二、三层为职工宿舍、阅览室、棋室、弹子房、会议室等，会议室的装修特别高雅，墙面用精美的红木镶板装饰，天花也用雕刻精美的红木花饰藻井。地下室为保险库、食堂等。该建筑系天津市文物保护单位。

上：三楼办公室走廊
Above: Third floor balcony colonnade

下：一楼外墙窗户
Below: Exterior of ground floor window

总经理室
The general manager's office

On May 27th 1905 the National Commercial Bank was founded by Zhejiang Railway Corporation, with its headquarters in Hangzhou. The total amount of equity fund was 1 million Yuan, only one fourth of which was original received. Hankou and Shanghai branches were set up in the same year. In 1915, Zhejiang Railway Corporation was nationalized with its shares taken over by private commercial capital, and therefore the National Commercial Bank was restructured and the Shanghai branch became the headquarters, therein the capital of 1 million Yuan was supplemented. On October 24th of

营业大厅
Business hall

the same year saw the founding of the Tianjin sub-branch, located in Gong Bei Steet. In 1921, the Tianjin sub-branch was upgraded to the Tianjin Branch, and moved to Rue de Chaylard in the French Settlement (now the intersection of Heping Road and Binjiang Road). In December 1952, the bank was nationalized by the communist government.

National Commercial Bank possessed the right (abolished in 1935) to issue exchange certificates. In 1923 Tianjin Branch of National Commercial Bank issued banknotes imprinted with the local name Tianjin, which covered 3 versions and denominated respectively by one Yuan, five Yuan, and ten Yuan.

National Commercial Bank, Zhejiang Industrial Bank, Shanghai Commercial Savings Bank, and Xin Hua Trust & Savings Bank Limited were together called the "South Four". Besides normal commercial bank business, National Commercial Bank concurrently handled agency business, leasing and warehousing business. Tianjin Branch's mainstream business was deposit-accepting, loan-making, currency exchange, and concurrently, it handled the trades of currencies as well as warehousing business.

Sited on No.319 Heping Road, designed by Shen Liyuan of Hua Xin engineering company, the building of the bank was constructed in 1922, which was of 4-story mixed structure with a semi basement. The building covers an area of 3,133 m² and a building area of 2,034 m². The facade is of classical design; the ground floor appears as the plinth, the second and the third floor as columns, the fourth floor as architrave, frieze, and cornice. Above the entrance there is an arc-shaped double column with a frieze, below there are Tuscan double columns shoring up the porch. The ground floor wall consists of deep jointed stone blocks, with semicircular arched windows, the key stone featuring a lion's head. The business hall is circular surrounded by 14 dark marble columns, on top of which rests a ring inlaid with a design of ancient Chinese coins. The business counter below the ring is sculpted with fine marble, again with a lion head motif. The hall is topped with a steel and ground glass dome producing a bright and airy interior. To the sides of the hall reside a manager's room, reception room, secretary's room and accountant's room. On the second and third floor are the employees' dormitory, reading room, chess room, billiard room and meeting room. The decoration in the meeting room is very graceful, the walls have red wood paneling and the ceiling is delicately-sculpted with red wood flower ornaments. The bank vault and dining hall are in the basement. The building classed as a Tianjin cultural relic. Now the building is used for commercial purposes.

银行保管库
Bank vault

上：通往保管库与金库的大门
Above: Bank vault door
下：大厅内穹顶下檐古钱币图案
Below: Dome decorated with a pattern of ancient coins

大厅中央的大理石喷泉
The marble fountain in the middle of the grand hall

营业大厅入口
Entrance to the business hall

华丽的总经理室
The magnificent general manager's office

会客室
Reception room

保管庫
Bank vault

YOKOHAMA SPECIE BANK, TIANJIN BRANCH

横滨正金银行

天津分行

昔日的横滨正金银行天津分行（今日中国银行天津分行）
The former Yokohama Specie Bank (Now Bank of China, Tianjin Branch)

金色的铜大门及窗间的铜饰板是财富的象征
The golden copper gate is a symbol of fortune

横滨正金银行
天津分行
YOKOHAMA SPECIE BANK, TIANJIN BRANCH

办公楼走廊
Corridor

日本横滨正金银行于1880年创办，总行设于日本横滨。清光绪二十五年（1899），横滨正金银行天津分行在英租界中街开业，光绪二十七年（1901）迁至英租界中街2号（今解放北路80号），1926年在该地建新楼营业。横滨正金银行曾在天津设立多家分店。1941年在日租界宫岛街（今鞍山道）成立分店，并将日本银行（中央金库）代理店一并移至分店。1945年7月31日日本储蓄银行天津分行所有资产负债并入横滨正金银行天津分行。横滨正金银行天津分行其原始资金仅300万日元，后期竟增加到1亿日元。该行自开设至1937年，名义上以国际汇兑为主要业务，实际上是依靠清朝政府和北洋政府的借款关系以及大量吸收清朝贵族、北洋军阀的"保价存款"而发达起来。该行曾扶植日本的洋行、会社开展在华业务，支持在天津的日本军工及特务活动，为日本侵华战争提供了雄厚的财力。1937年七七事变后，该行受日本政府指令扶植成立伪中国联合准备银行。1945年8月日本投降后，该行由中国银行天津分行接收。

横滨正金银行天津分行在1902~1937年期间曾发行大量纸币，分银两和银元兑换券两种。其中发行加印横滨正金银行天津地名纸币，银两券有5两、10两、50两和100两，银元券有1元、5元、10元和100元。

横滨正金银行天津分行下设支配人席（经理）、秘书课、预金课（存款）、送金课、输出入课（只办理对华中、华南交易）、贷付课、电信课、计算课、庶务课、考查课和华账房，主要经营代理日本银行业务、存款、放款、汇款、国际贸易结算和代保管等业务。1937年上半年赢利额为30.2万元。

横滨正金银行天津分行旧址位于解放北路80号，建于1926年10月，由英商阿特金森(Atrinson)和达拉斯(Dallas)经营的同和工程司设计，设计者为该公司的苏格兰建筑师伯内特(B.C.G Burnett)。由华胜建筑公司施工。占地面面积2 830m²，建筑面积3 150m²，是希腊古典复兴形式，古典主义三段论明确，建筑构图严谨。正面是八根科林斯巨柱式柱廊，两端用壁柱收尾，正立面上下两层窗间墙用金色花铜板装饰，整个檐部高与柱身比约为1:4。入口设首层，中央为金色花铜板装饰大门，花岗岩墙面，建筑整体稳重、华丽。建筑平面为矩形，功能分区明确，首层是300m²的营业大厅。大厅中央顶部为双层玻璃顶棚，下层为九格井字梁，嵌彩色玻璃，既美观又可为大厅采光。大厅顶部有机械通风设备，利用营业柜台做管道，将地下室通过天然冰冷却的空气排入大厅，这在当时是非常先进的。该建筑系天津市文物保护单位，特殊保护等级历史风貌建筑。现为中国银行天津分行。

上：银行办公大厅天顶
Above: Ceiling of the business hall

下：大厅西侧支撑天顶的两根爱奥尼克式巨柱
Below: Two huge Ionic columns support the ceiling on the west side of the hall

Japanese Yokohama Specie Bank Limited was founded in 1880 with its headquarter in Yokohama, Japan. In 1899, Tianjin branch of the Yokohama Specie Bank Limited opened on Victoria Road in the British Settlement, and in 1901, it moved to No.2 Victoria Road (now No.80 Jiefang North Road). In 1926, a new building was constructed at the original site. The Yokohama Specie Bank Limited set up many branches in Tianjin. In 1941, Yokohama Specie Bank Limited established a branch in Miyajima Road in the Japanese Settlement (now An Shan Road), this branch conducted business for Bank of Japan by proxy. On July 31st 1945, the assets and liabilities of Tianjin branch of Savings Bank of Japan were all added to that of Tianjin branch of the Yokohama Specie Bank Limited. Its original equity fund was only 3 million Japanese yen,

营业大厅入口处旋转门
Revolving door leading to the business hall

营业大厅
Business hall

通往营业大厅的走廊
Corridor leading to the business hall

营业大厅
Business hall

上：二楼会客室
Above: Second floor reception room

下：大理石与实木相结合的楼梯间
Below: Marble and wooden staircase

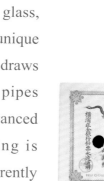

横滨正金银行在天津发行的银票
Banknotes issues by Yokohama Specie Bank in Tianjin

and later increased 100 million yen. From its establishment to 1937, its main business was international exchange of currencies. In fact its prosperity depended not only on loaning to the Qing government and the Northern Warlords Government, but also in dealing in vast quantity of the so called "inflation-proof deposits" of Qing Dynasty nobles and northern warlords. The bank once helped Japanese firms and corporations develop their business in China and supported Japanese military and spy activities, which provided huge financial resources for the war of invading China. After the July 7th incident of 1937, the bank propped up the China Joint Reserve Bank at the behest of the Japanese government. After the Japanese surrender in August 1945, Tianjin branch of Yokohama Specie Bank Limited was taken over by the Bank of China, Tianjin branch.

From 1902 to 1937 Tianjin branch of the Yokohama Specie Bank Limited had issued a great deal of banknotes divided into exchange certificates of silver and that of silver dollar, and some of which were printed with the local name Tianjin. Denominations of the exchange certificates of silver were 5 liang, 10 liang, 50 liang and 100 liang; and denominations of the silver dollar certificates were 1yuan, 5 Yuan, 10 Yuan, 100 Yuan.

Tianjin branch of the Yokohama Specie Bank Ltd. set up various divisions; manager office, secretary office, savings division, remittance division, import-export division, loan-making division, telecommunication division, settlement division, general affairs division, auditing division and brokerage division. The profit of the first half of 1937 was 302,000 Yuan.

Sited on the No.80 Jiefang North Road, the building was designed in 1926 by the Scottish architect B.C.G Burnett working for the British Atkinson and Dallas engineering company. Construction was by Hua Sheng Building Construction Co. Ltd., covering an area of 2,830 m² and building area of 3,150 m². The building is of Greek classical revival in form. The architecture composition is compact with classical proportions, the facade forms 8 huge Corinthian columns. The windows and entrance are adorned with golden flower ornamental copper panels. The walls are of granite adding a sense of stability and magnificence to the building. The plan is rectangle, with a 300 m² business hall on the first floor. The hall is topped with a double layered glass ceiling, below which there are 9 beams inlaid with stained glass, creating a stunning and light room. A unique feature is the ventilating system, which draws cool air from the basement through pipes in the counters; this was the most advanced technology of the day. The building is protected as a Tianjin cultural relic. Currently it is occupied by the Bank of China, Tianjin Branch.

会客室壁炉
Reception room fueplace

营业大厅二楼走廊
Second floor corridor off the business hall

会客室
Reception room

INDUSTRIAL AND COMMERCIAL BANK OF CHINA-FRANCE, TIANJIN BRANCH

中法工商银行

天津分行

原中法工商银行天津分行（今日天津市总工会）
The former Industrial and Commercial Bank of China-France, Tianjin Branch (Now the Tianjin Federation of Trade Unions)

科林斯巨柱与弧形柱廊
The huge Corinthian columns and arched colonnade

大堂内的罗马柱
Roman column inside grand hall

中法工商银行
天津分行

INDUSTRIAL AND COMMERCIAL BANK OF CHINA-FRANCE, TIANJIN BRANCH

中法工商银行前身为中法实业银行，成立于1913年1月，为中法合资的股份有限公司组织，根据法国的法律办理注册，并取得中国政府准许。总行设于巴黎。资本总额定为4 500万法郎。中国政府认购该行股票3万股，计1 500万法郎，由法方垫付，实权由法方掌管。中法工商银行天津分行于1916年11月开业。1921年因故停业。1923年11月复业更名为中法工商银行，继续营业，1948年底停业。

中法实业银行成立后即取得在中国纸币发行权。1916年中法工商银行天津分行发行加印中法实业银行天津地名纸币有1元、5元、10元、50元、100元和500元共计6个版别。

中法工商银行天津分行主要业务范围有：存款、买卖外汇和股票、汇款、借贷、代销法国彩票、有奖债券、发行钞票、企业投资等。

中法工商银行天津分行旧址在法租界中街西宾馆，即今解放北路74号。由法商永和工程司(Brossand mopin)的法国建筑师马利奎特(Maliquet)设计，建于1933年，建筑风格为罗马古典复兴式。主楼为四层混合结构带半地下室，占地面积1 567m²。建筑面积6 240m²，建筑物入口在马路转角处，有10根科林斯巨柱式弧形空柱廊，空柱廊两侧为实墙，开有门窗，沿解放北路立面为4根科林斯壁柱，使立面主次分明。三层及顶层双柱空柱廊是后建的。顶层双柱空柱廊有瓶饰栏杆。主要入口柱廊上方的中央三跨双柱空廊在三层上适当地挑出并用8对檐托支撑，使巨柱式空柱廊入口更为明显。营业大厅面积213m²，大厅内采用黑白相间马赛克地面和仿石砌墙壁。柜台内侧采用罗马陶立克圆柱6根，方柱2根，顶部装有彩色玻璃采光窗。该建筑系天津市文物保护单位，特殊保护等级历史风貌建筑。

建筑外墙局部
Bank exterior

建筑外部花饰
Bank exterior

二楼办公室走廊
Second floor offices corridor

Entrance hall

Above: Business hall entrance hall

Below: Office fireplace

The Industrial Bank of China-France was founded in January 1913. This Sino-French joint-venture had its headquarters in Paris and was registered in accordance with French laws. The total equity fund was 45 million Francs. The Chinese government subscribed to 30 thousand shares, worth 15 million Francs. The Tianjin branch opened in November 1915 and ceased business for some reason in 1921. However, it resumed business and was renamed as the Industrial and Commercial Bank of China-France in November 1925. When Tianjin was under Japanese occupation, the bank was fortunate to avoid being taken over and kept doing business. By the end of 1948, it terminated its business and closed.

Industrial and Commercial Bank of China-France acquired the currency issuing right shortly after its founding. In 1916, Tianjin branch issued banknotes printed with the name Tianjin, included 6 denominations the 1 Yuan, 5 Yuan, 10 Yuan, 50 Yuan, 100 Yuan and 500 Yuan.

Industrial and Commercial Bank of China-France's mainstream business covered deposit-accepting, foreign exchange and securities trading, remittance, borrowing and lending, proxy sale of French lottery, premium bonds, banknote issuing and enterprise investment, etc.

Located in the West Hotel on Rue de France in the French Settlement (now 74 Jiefang North Road), the building itself was designed by Maliquet, French architect of Brossand Mopin and constructed in 1933. The building is in the form of Roman Classical Revival. The main building is of 4-stories plus semi basement with an area of 1,567 m² and floor space of 6,240 m². The entrance is on a corner with 10 huge Corinthian columns in an arc. The double-column colonnade on the third and fourth was a later addition. The 213 m² business hall features a black and white stripped mosaic floor and a stone wall. Behind the counters, there are 6 Roman Doric columns and 2 pillars. The ceiling is of stained glass. The building is protected as a Tianjin cultural relic.

Delicately carved stone flower

Finely carved Corinthian columns

办公室走廊
Offices corridor

营业大厅
Business hall

会客室
Reception room

透过略有雕琢的铁窗，蓦然看到窗外的繁华
Busy street viewed through ironwork window

CENTRAL BANK,
TIANJIN BRANCH

中央银行

天津分行

昔日的中央银行天津分行（今日的中国人民银行天津分行）
The former Central Bank, Tianjin branch (Now the Peoples Bank of China, Tianjin Branch)

中央银行
天津分行
CENTRAL BANK, TIANJIN BRANCH

中央银行发行的纸币
Banknotes issued by Central Bank

国民政府中央银行于1928年11月1日在南京正式成立。1931年4月10日中央银行天津分行成立并开业，行址在英租界中街（抗战胜利后改称中正路，今解放北路）。1935年天津分行定为一等分行。1937年"七七"事变后，天津分行撤退。1945年抗战胜利后，中央银行天津分行于原址复业。1949年1月15日天津解放后，该行被天津市军事管制委员会接管部接收。

中央银行天津分行成立后，根据国民政府颁布的银行法规定，行使中央银行职能，办理中央银行各种兑换业务及中央银行各种兑换券的发行等各项业务。1934年中央银行天津分行曾在津发行加印天津地名纸币。1935年11月国民政府实行法币政策后，该行负责接收天津各发行银行所发兑换券的现金准备、保证准备及未收回的新、旧钞票，并继续办理收兑。截至1937年11月，中央银行天津分行发行总额为3 234.3万元。1946年底，中央银行天津分行存款余额5 465 701万元，放款余额416 090万元，纯收益为393 175万元。

中央银行天津分行旧址位于今解放北路117号，原为中日合资的中华汇业银行，由华信工程司沈理源设计，建于1925年，1926年建成。该建筑为三层带半地下室混合结构，建筑面积4 245m²。建筑立面为古典复兴式，古典主义三段论设计很明确。半地下室作为台基，底层左侧开有券洞式旁门，为进入楼内的通道。一、二层用4根巨柱式爱奥尼克空柱廊统一起来，三层作为檐部加阁楼层的形式出现，檐口出檐很小，较简洁。中部为波浪形山花，强化垂直轴线布局。入口由高台阶上门厅，再进入营业大厅，大厅简洁的玻璃顶受维也纳学派的影响。经楼梯上二层，大厅用类似爱奥尼克柱子支撑，顶棚是色彩鲜艳的小八角形藻井，甚为华丽。二层为办公室，会议室等。会议室内部装修极为讲究，全部用高级硬木雕饰，室内用类似爱奥尼克木雕圆柱支撑，圆柱旁有木雕方壁柱，雕刻精细。墙面用高级硬木护墙板，顶棚周边用齿饰、剑蛋饰、檐托，衬托出雕刻精细的木藻井。三层为职工宿舍等，砖混结构。该建筑系天津市文物保护单位，特殊保护等级历史风貌建筑。现为中国人民银行天津分行。

办公楼入口处
Office building entrance

大厅天顶
Ceiling of the grand hall

The Central Bank of the National (KMT) government was founded in Nanjing on November 1st, 1928. Tianjin Branch of the Central Bank was founded and opened on April 10th, 1931 on Victoria Road in the British Settlement (now Jiefang North Road). In 1935, the Central Bank prescribed the first, second and third-grade branches with Tianjin Branch set as first-grade. Tianjin Branch was closed after the Marco Polo Bridge Incident of July 7th 1937. After the victory in the Anti-Japanese War in 1945, the Tianjin Branch of the Central Bank resumed its business at the former address. After Tianjin's jiefang on January 15th, 1949, the branch was taken over by the Tianjin Military

大楼外景
Bank exterior

营业大厅
Business hall

二楼会议室
Second floor meeting room

上：二楼办公室走廊
Above: Second floor corridor

下：楼梯间
Below: Staircase

金库大门局部
Part of the vault door

Control Committee.

After its founding, Tianjin Branch of the Central Bank functioned as the central bank to handle various kinds of exchange business and the issue of exchange certificates. In 1934, Tianjin Branch of the Central Bank issued paper money printed with Tianjin place-names. After the national government adopted the policy of issuing fiat money in November, 1935, the branch was obliged, in addition to keeping on exchange business, to take over the cash reserves and reserve requirements for various exchange certificates from the issuing banks in Tianjin and recalled old banknotes. Up to November 1937, Tianjin Branch of the Central Bank's issuing amount had totaled 323.43 million Yuan. At the end of 1946, it had a deposit balance of 54.65701 billion Yuan, loan balance of 4.1609 billion Yuan, and the net return of 3.93175 billion Yuan.

Sited on No. 117 Jiefang North Road, the building of Tianjin Branch of the Central Bank was designed by Shen Liyuan of Hua Xin Engineering Company and constructed in 1925. The buildings elevation is of classical style. The arched side door at the left side of the first floor leads is the entrance to the building. The first and the second floor are unified by 4 huge Ionic colonnades. The wave-shaped pediment reinforces the vertical axis. Steps lead from the entrance way to the business hall. The glass ceiling is of Viennese design. Through the staircase up to the second floor, there is a hall supported by Ionic columns. On the top, there is a magnificent octagonal recessed ceiling. On the second floor are offices and meeting rooms. The interior decoration of meeting room is very exquisite all using high quality hard wood sculptural details. The room is supported by wooden Ionic columns and delicate wooden pilasters. High quality wood paneling cover both walls and ceiling, with egg and dart motifs, eaves struts are used to set off the finely carved wooden ceiling. The third floor made of brick and concrete houses the employees' dormitory. The building is protected as a Tianjin cultural relic. It is now occupied by the People's Bank of China, Tianjin Branch.

二楼会客室
Second floor reception room

镜中反射的护墙板和家俱
Reflection in a mirror

会客室一角
Reception room

二楼办公室前厅用类似爱奥尼克的柱子支撑，顶棚是色彩鲜艳的小八角形藻井
Lonic columns support the entrance hall of the second floor office, on top is the colorful caisson ceiling

RUSSO-CHINESE BANK, TIANJIN BRANCH

华俄道胜银行

天津分行

Decorated arched window, Russian-style curved cornice

华俄道胜银行
天津分行
RUSSO-CHINESE BANK, TIANJIN BRANCH

华俄道胜银行于1895年（清光绪二十一年）创办，开始由俄法两国合资，资本600万卢布。次年为中俄合资，清政府从俄法借款项下拨出500万两白银作为投资。总行设在彼得堡。1910年（清宣统二年）与另一家俄法合资的银行——北方银行合并，改称俄亚银行。中文名称不变。十月革命后总行被苏维埃政权收归国有，总行改设巴黎，并继续在中国经营。1926年巴黎总行因外汇投机失败而清理倒闭。清光绪二十二年（1896年）华俄道胜银行天津分行开业，地址在英租界中街（今解放北路）与领事道（今大同道）的转角处。光绪二十六年（1900年）在原地重新建楼。1926年随在华各地分行一起倒闭。

华俄道胜银行天津分行营业期间，根据沙俄政府颁布的《华俄道胜银行条例》规定，银行有权代收中国各种税收，有权经营与地方及国库有关的业务，可以铸造中国政府许可的货币，代还中国政府所募公债利息，敷设中国境内铁道和电线等项工程。该行在华还大量发行纸币，代收税款，向旧中国政府提供政治贷款和铁路贷款等。

华俄道胜银行天津分行发行的加印华俄道胜银行天津地名纸币分银两和银元兑换券两种，银两券有1两、3两、5两、10两、50两、100两和500两，银元券有1元、5元、10元。

华俄道胜银行天津分行大楼由德国建筑师查理·西尔(Richard Seel)设计，后经几次改建，银行入口处在马路转角处，该建筑采用了文艺复兴时期的穹顶及采光亭，又采用了罗马时期的圆拱券，还采用了巴洛克时期的曲线形的尖山墙，是各种风格的综合，属于折衷主义建筑形式。自入口进入六角形的门厅，经门厅内两侧的弧形台阶进入底层营业大厅。大厅为对称的短"L"形，两侧为办公室、接待室。二层为职工卧室，餐厅及会客室、音乐厅。三层为勤杂人员卧室、厨房、洗衣房等，半地下室为金库、账库等。该建筑为天津唯一用穹顶、采光亭的建筑。现在系天津市文物保护单位，特殊保护等级历史风貌建筑。现由中国人民银行天津分行使用。

上：办公楼大门
Above: Office building entrance

下：穹顶内部结构
Below: Inner surface of the dome

从横滨正金银行看华俄道胜银行
View of Russo-Chinese Bank, Tianjin Branch from Yokohama Specie Bank.

Russo-Chinese Bank was founded in 1895, with initial investment from Russia and France with a total capital of 6 million ruble. The next year the Qing Dynasty allocated 5 million silver liang out of the loan from Russia and France as the investment, and subsequently the bank become a China-Russia joint venture, with its headquarters in St. Petersburg. Then in 1910 it merged with the Northern Bank---another Russia-France joint-venture bank, and was renamed as Russo Asian Bank (the Chinese name was unchanged). Because the building of the headquarters was nationalized by the Soviet regime after the October Revolution, the headquarters had to move to Paris, and the bank kept continued business in China. The headquarters in Paris went bankrupt as a result of the failure of foreign exchange speculations. Russo Asian Tianjin opened in 1896 on the corner of Victoria Road (now Jiefang North Road) and the Consul Street (now Datong Road) in the British Settlement. The Tianjin branch of Russo-Chinese Bank bankrupted along with other branches in China in 1926.

银行营业大厅
Business hall

曾经的穿衣镜
Original dressing table

曾经的老家具
Original furniture

The Russo Asian Bank was entitled to collect all kinds of revenues in China on behalf of Russian government; to manage business related to local affairs and treasury; to mint coins under the permission of the Chinese government; to represent Chinese government to pay the interest of government bonds, to lay railway tracks and electric wires within China's boundaries, etc. This bank also had issued paper currency in great quantity, acted on collecting taxes and provided political/railway loans for Chinese government, etc.

The banknotes printed with the name Tianjin issued by the Russo Asian Tianjin were divided into exchange certificates of silver and of silver dollar. Silver certificates were denominated as one liang, three liang, five liang, ten liang, fifty liang, one hundred liang and five hundred liang. Denominations of silver dollar certificates were one dollar five dollar, and ten dollar.

The original designer was the German Richard Seel. It was rebuilt several times later. The bank entrance is on a corner. The building adopts the dome and lantern in Renaissance period as well as Roman circular arch. It also adopts curved pointed pediment of Baroque period, which is a mix of all styles. The architecture is an eclectic style. The entrance leads to the hexagonal hall, by way of curving steps each side of the hall. The business hall is on the ground floor. The hall is in a short "L" shape, with offices and reception room on two sides. On the second floor there are the employees' bedrooms, dining hall, reception room and music hall. On the third floor are the servants quarters, kitchen, laundry house, etc. In the basement is the exchequer, account book house, etc. This building is the only one in Tianjin to adopt a dome and lantern. The building is protected as a Tianjin cultural relic. Currently it is occupied by the People's Bank of China, Tianjin Branch.

华俄道胜银行夜色
Night view of Russo-Chinese Bank

办公室楼梯间
The office staircase

营业大厅入口过厅
Entrance hall of the business hall

KINCHEN BANK

金城银行

昔日的金城银行
The former Kinchen Bank

金库大门的保险锁
The security lock for the gate of safe

金城银行
KINCHEN BANK

金城银行为金融界知名人士周作民于1917年5月15日创办,其主要股东多为军阀官僚。"名曰金城,盖取金城汤池永久坚固之意"。总行设于天津,行址在法租界7号路(亦称法租界中街,今解放北路43号),1921年2月迁英租界中街(今解放北路108号)。金城银行是中国重要的私营银行之一,是"北四行"(金城银行、盐业银行、中南银行、大陆银行)的主要支柱。该行的股本资金主要来自北洋军阀和官僚,开办时原定资本200万元,实收50万元,至1919年收足,并增资到500万元,1923年再次增资为1 000万元,截至1927年4月实收700万元。1936年金城银行总行迁往上海,天津总行改为分行。1952年12月该行参加全行业公私合营,与其他银行、钱庄一起组成统一的公私合营银行。

1917~1935年期间,金城银行曾先后在北平、上海、汉口、大连、哈尔滨、南京、青岛等地设有分行,在蚌埠、包头、张家口、绥远、武昌、苏州、郑州、常熟、长沙、定县、潼关、许昌、新乡等地设有办事处。

金城银行积极开展储蓄业务,大力吸收教会、团体、教育界、医药界的存款以及社会中立阶级的闲散资金。1936年的存款达1.83亿元,一度超过上海商业储蓄银行,居私营银行首位。

金城银行放款的半数以上集中于少数重点客户,工业放款中主要投向棉纺织、化学、煤炭和面粉工业。金城银行曾扶植著名的天津永利公司"红三角"牌纯碱打入国际市场。金城银行投资业务也很发达,其中直接设立自营的附属企业太平保险公司,后来成为华商保险业的"巨擘"。

金城银行旧址位于解放北路108号,由德国建筑师贝克、培迪克(Bercker&Baedecker)设计,建于1907~1908年。由汉堡市阿尔托纳区F.H.施密特公司(F.H.Schmidt.Atona,Hamburg)施工,建筑面积5 933m²,为砖混结构2层带阁楼层西式楼房。该建筑采用高级复折四坡屋顶的混合结构。底层立面中央用4根粗壮的立柱支撑,入口大门处用2根类似罗马多立克柱子支撑一个短的檐部。檐部上面有两个牛腿支撑着二层出挑的阳台。二层中央立面用四对带有中国回纹柱头的双柱支撑,入口上部用2根粗壮的牛腿与其他4对双柱共同支撑着二层的檐部。二层出挑阳台的铁栏杆花纹十分精细,除回纹外还有略似洛可可式的花纹。屋顶除高坡复折四坡屋顶外入口处有山墙及3个老虎窗。始建时为德华银行天津分行(German Asian Bank,Tianjin Branck)后改为金城银行所用。1937年由华信工程司沈理源对银行内部适当地加以改造。内部装饰精美,楼梯间的木栏杆雕刻精细,楼梯窗户的彩色玻璃色彩鲜艳。由于当时上海、天津很多银行都采用折衷主义的形式,所以也采用了回纹。设计精巧,建筑精美,别有风采。该建筑系天津市文物保护单位,重点保护等级历史风貌建筑。

后楼门窗
Rear facing window

镶有铁艺护栏的窗户
Ironwork window

银行外围墙
Bank exterior

Kinchen Bank was founded on May 15th, 1917, by Zhou Zuomin, a prestigious figure in financial field then. Its substantial shareholders were mostly warlords and bureaucrats. "The name Kinchen literally in Chinese means gold city, symbolizing permanence and firmness". Headquartered in Tianjin, the bank was originally sited at Rue de France in the French Settlement (now No.43 Jiefang North Road), and then moved to Victoria Road of the British Settlement in 1921 (now No.108 Jiefang North Road). Kinchen Bank was one of the most important private banks and the pillar of the "North Four" (Kinchen Bank, the Salt Industrial Bank, the China & South Sea Bank Limited and Mainland Bank). With the equity fund mainly coming from the Northern warlords and bureaucrats, the amount of the originally agreed capital was 2 million Yuan at the

establishment, but the paid-up capital was only 0.5 million Yuan. By 1919, the rest had been paid up and the total capital had risen to 5 million Yuan, and in 1923, it was raised again to 10 million Yuan. By the end of April 1927, the amount of paid-up capital had risen to 7 million Yuan. In 1936, Kinchen Bank moved its headquarters to Shanghai and the former Tianjin headquarters became a branch. In December 1952, the bank was nationalized by the communist government.

From 1917 to 1935, Kinchen Bank successively set up branches in many cities including Peking, Shanghai, Hankou, Dalian, Harbin, Nanjing, Qingdao and representative offices in secondary cities such as Bengfu, Baotou, Zhang Jiakou, Suiyuan, Wuchang, Suzhou, Zhengzhou, Changshu, Changsha, Dingxian County, Tongguan, Xuchang, Xinxiang.

Kinchen Bank actively developed its savings business, vigorously absorbed deposits from churches, organizations, educational and medical fields. In 1936, the deposits amount had reached 183 million Yuan, which once exceeded that of Shanghai Commercial Savings Bank and ranked first among private Chinese banks.

More than half of Kinchen Bank's loans focused on a few VIP clients, industrial loans to the textile, chemical, coal mining and flour milling industries. It assisted the "Red Triangle" soda ash of Tianjin Yongli Chemical Industries into the international market. Kinchen Bank's investment business was also very prosperous. It directly established a self-support subsidiary—Taiping Insurance Company, which later became a major insurance provider in China.

Located at No.108 Jiefang North Road, the building of Kinchen Bank was designed by German architects Bercker & Baedecker, and constructed by F.H.Schmidt, Altona, from Hamburg in 1907 to 1908. It is a 2-story brick and concrete structure with a high pitched roof containing attic rooms, with a total floor space of 5,933 m². The central of the ground floor elevation is supported by 4 pillars. At the entrance, there are 2 similar Roman Doric columns. The central of the second floor is supported by 4 pairs of columns with capitals displaying a Chinese-style winding pattern. The second floor has a Rococo style balcony with ironwork depicting a delicate flower pattern. There are 3 dormer windows in the roof. The building was originally occupied by the Tianjin branch of German Asian Bank, later it was acquired by Kin Cheng Bank. In 1937, Shen Liyuan of Hua Xin Engineering Company remodeled the interior decoration of the bank. The interior ornaments are fabulous; the wooden banister on the staircase is delicately carved; the stained glass of the stairway windows is beautifully colored. The Chinese style winding pattern is very typical of many banks in Shanghai, Tianjin at that time. Ingenious design and fabulous architecture make it very unique. The building is protected as a Tianjin cultural relic.

总经理办公室
The general manager's office

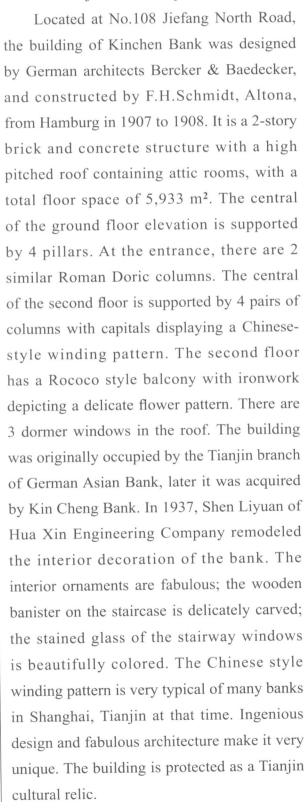

办公楼入口处
Office building entrance

上：保管库
Above: The safe

下：护墙板一角
Below: Paneling

办公室楼梯间
Office staircase

办公室走廊
Offices corridor

站在总经理室门前俯视金融街
View from the manager's office of the financial street

金城银行夜景
Night view of Kinchen Bank

THE CHARTERED BANK OF INDIA, AUSTRALIA & CHINA, TIANJIN BRANCH

麦加利银行

天津分行

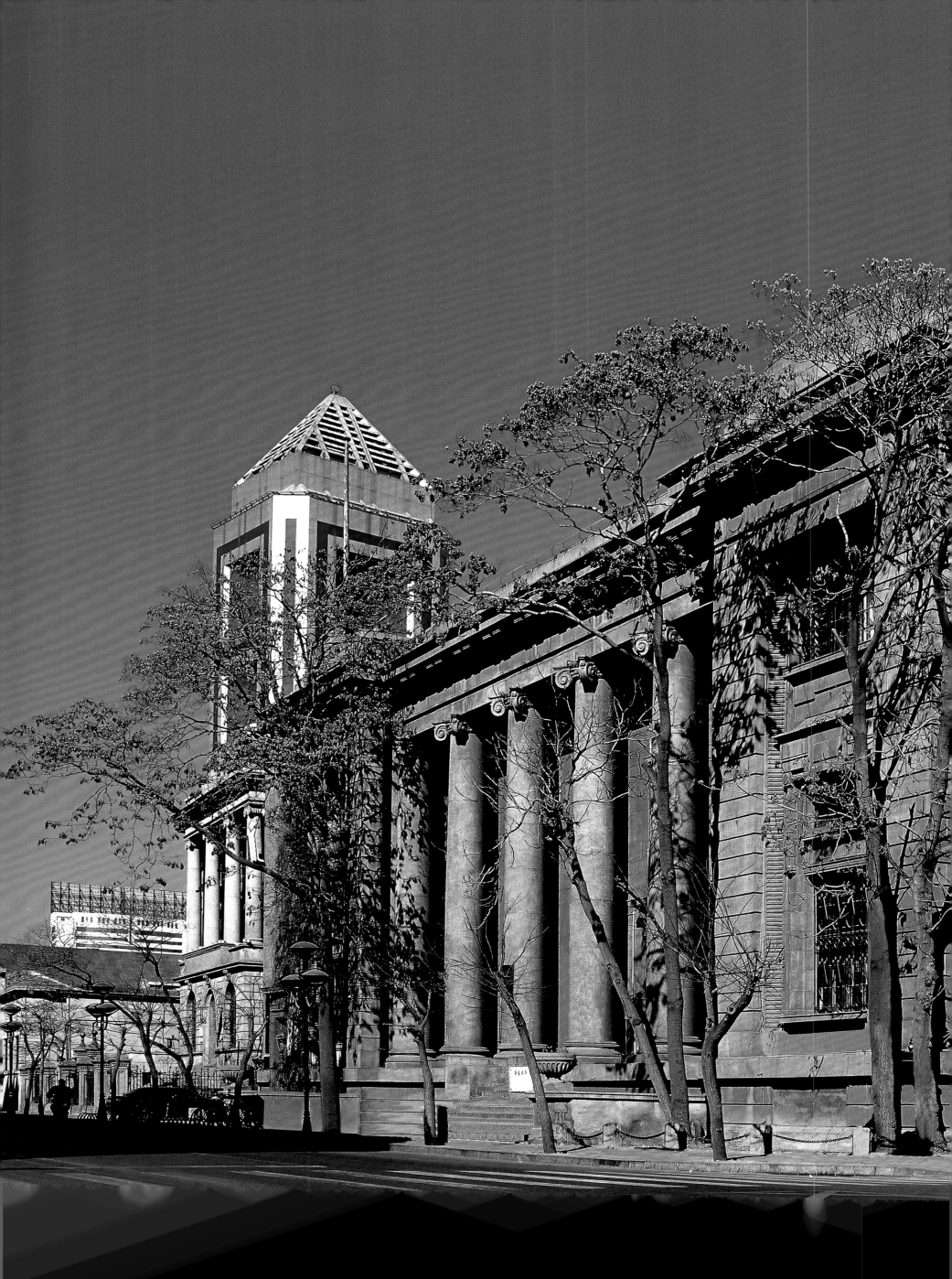

麦加利银行外景
Babk exterior

旋转门上的木雕花饰局部
Part of the carved wooden flower on the revolving door

麦加利银行
天津分行

THE CHARTERED BANK OF INDIA, AUSTRALIA & CHINA, TIANJIN BRANCH

麦加利银行在天津发行的纸币
Banknotes issued by Tianjin Branch of the Chartered Bank of India, Australia & China in Tianjin

英国麦加利银行亦称渣打银行，创办于1853年，总行设于伦敦。清光绪二十一年（1895年）麦加利银行天津分行在英租界中街79号（今解放北路153号）开业，1925年在原址建成新楼。1941年太平洋战争爆发后，麦加利银行天津分行被日军接管。1945年抗战胜利后，该行在原址恢复营业。1954年清理歇业。

麦加利银行天津分行发行的钞票分银两和银元兑换券两种，银两券有5两、10两、50两和100两，银元券有1元、5元、10元、25元、50元、100元和500元。其中发行加印麦加利银行天津地名纸币有5元、10元、25元、100元和500元。1933年麦加利银行天津分行发行钞票达91.4万元。

麦加利银行天津分行经营范围甚广，下设出口部、入口部、电汇部、汇兑部、出纳部和流水账部等，主要办理定活期存款、放款、汇兑信用证、外汇买卖业务等。另设保管部和信托部，代客保管一切物件、募集债券、发行和买卖证券及其他受托事项。1933年麦加利银行天津分行资本、公积金和各项存款总计约占总行的7%，达到347.4万元。

麦加利银行天津分行旧址位于解放北路153号，1924年由景明工程司英国人赫明（Hemming）和伯克利（Berkley）设计，1925年建成。该建筑为钢筋混凝土框架结构，二层带半地下室。占地面积2 259m²，建筑面积5 933m²。平面布局近平形四边形，西洋古典风格，整体建筑宏伟庄重。大楼基座用花岗岩砌筑，入口在解放北路，立面为五跨4根巨柱式爱奥尼克空柱廊，其中两侧呈3/4圆柱。柱头圆形卷涡呈45角斜角式，入口柱廊两侧为长方形水泥砌块。太原道沿街里面也为五跨巨柱式爱奥尼克空柱廊，但用的是3/4圆的壁柱，整个立面古典主义三段论很明确，显得庄严肃穆。门外台阶两侧各有一个西式大花钵，花钵下的水泥柱墩，以粗铁索连接成栏杆，形成一个小空间，显得华丽典雅。营业大厅面积651.2m²，呈短"L"形布局，钢门窗、木制旋转门豪华大方。该建筑系天津市文物保护单位、特殊保护等级历史风貌建筑。

上：奢华的旋转门
Above: Revolving door

下：外檐的装饰
Below: Decorated eaves

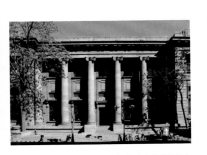

银行正门由4根高大的爱奥尼克巨柱组成的柱廊
The four huge Ionic column colonnade in front of the main entrance

British Chartered Bank of India, Australia & China, also named Chartered Bank, was founded in 1853 with its headquarters in London. In 1895 the Tianjin branch was opened at No. 79 Victoria Road in the British Settlement (now No.153 Jiefang North Road). In 1925 a new building was constructed at the original site. In 1941, when the Pacific war broke out, the Tianjin branch was taken over by the Japanese Army. After the victory in the Anti-Japanese War, the bank resumed its business at the original address. In 1954, it was closed.

The banknotes that Tianjin branch of the Chartered Bank of India, Australia & China had issued were classified into exchange certificates of silver and that of silver dollar. Silver certificates were divided into 5 liang, 10 liang, 50 liang and 100 liang certificates, while the denomination of the silver dollar certificates were respectively 1 dollar, 5 dollar, 10 dollar, 25 dollar, 50 dollar, 100 dollar and 500 dollar, all of which except 1 dollar note were printed with Tianjin places names. In 1933, the banknotes issuing

营业厅入口的旋转大门
Revolving door of the business hall

Offices in the business hall

amount of Tianjin branch reached 914,000 dollar.

Covering a wide range of businesses, the Tianjin branch of the Chartered Bank of India, Australia & China set up export department, import department, telegraphic transfer department, exchange department, receive and payout money department and journal account department. In addition, safe deposit department and trust department were set up to be in charge of keeping clients' items, collecting bonds, issuing, buying and selling securities as well as accepting other entrustments. In 1933, the total amount of the Tianjin branch's capital, funds and deposits accounted for 7% of that of the headquarters, reaching 3.474 million dollar.

The building of the Chartered bank, sited at the No.153 Jiefang North Road, and designed by Hemming and Berkley's Engineering Company. Constructed in 1925 it is a 2-story building of reinforced concrete frame structure covering an area of 2,259 m² and floor space of 5,933 m². The plan is in the shape of parallelogram. It is of western classical style, wholly grand and magnificent. The foundation is of solid granite. Its entrance on the Jiefang North Road has 4 huge Ionic columns. Each side of the columns, there are rectangular cement blocks. Each side of the steps outside the entrance there is a western style flower bowl. The business hall covers area of 651.2 m², with an L-shaped layout, steel doors and windows, and a luxurious wooden revolving door. The building is protected as a Tianjin cultural relic. Now the building serves as the Jiefang North Road Post Office.

Counter entrance

Office building lift

Business hall

营业大厅
Business Hall

THE NORTHERN FOUR SAVING BANKING SOCIETY, TIANJIN BRANCH

四行储蓄会

天津分会

昔日的四行储蓄会天津分会（今日的中国工商银行天津分行）
The former building of The Northern Four Saving Banking Society, Tianjin Branch (Now China Commercial and Industrial Bank, Tianjin Branch)

四行储蓄会
天津分会
THE NORTHERN FOUR SAVING BANKING SOCIETY, TIANJIN BRANCH

1923年6月中南银行、盐业银行、金城银行、大陆银行等四家银行联合成立四行储蓄会，总会设在上海。同时在天津、上海、南京、北平、汉口等地设有9个分会。四行储蓄会的"基本储金"（资本总额）100万元，四家银行各投资25万元，专门办理储蓄业务。

四行储蓄会储蓄业务分定期储金、分期储金、长期储金、特别储金、活期储金、保管公益款项、满期储金等种类。定期储金利率规定为年息7厘，每半年办理决算一次。分配盈余，先提基本酬金（即银行担保本息的酬金），再提公积金、会员红利（即给存款人的红利）及职员酬劳金。利息7厘加会员红利合计最高为1分1厘7，最低为8厘7。此举措深受存款人欢迎。

四行储蓄会开业后储蓄存款逐年增长。1923~1936年，四行储蓄会各项储金总额达9 000余万元，成为当时全国最大的储蓄银行。该行储蓄存款主要用于工商业抵押放款、购买公债和各种有价证券及存放四银行和同业拆借。1937年太平洋战争爆发后，该行业务逐渐萎缩。

四行储蓄会天津分会至今传颂着一段爱国主义的佳话。1932~1949年盐业银行天津分行经理陈亦侯和四行储蓄会天津分会经理胡仲文曾将末代皇帝溥仪向盐业银行借款抵押的16只金编钟藏在四行储蓄会天津分会大楼的地下库房里。先后躲过了军阀、日本军警和国民政府的搜查掠夺。1949年1月18日，天津解放的第三天，藏匿多年的金编钟终于重见天日，全部交给国家，现陈列在故宫博物院珍宝馆。

四行储蓄会天津分会旧址位于解放北路147号，建于1923年。建筑面积1 590m²，为砖混结构3层建筑，平面布局呈三角形。首层立面有1个拱券门2个拱券窗，二层分立4根爱奥尼克立柱，顶部出檐，设计简洁，装修精美。该建筑系天津市文物保护单位，特殊保护等级历史风貌建筑。现为中国工商银行天津分行。

上：充满古典风格的窗户
Above: Classic-style window

下：入口大门
Below: Main entrance

大楼外景
Bank exterior

In June, 1923, the China & South Sea Bank Ltd, Salt Industrial Bank, Kinchen Bank and Mainland Bank jointly founded The Northern Four Saving Banking Society headquartered in Shanghai. At the same time 9 branches were set up in Tianjin, Shanghai, Nanjing, Peking, Hankou, etc. the Joint Savings Society Bank's "savings base" was 1 million Yuan with each of the four investing 250,000 Yuan to specifically to handle savings business.

The Northern Four Saving Banking Society classified its savings business into time deposit, time deposit of small savings for lump-sum withdrawal, long-term deposit, special savings, demand deposit, contribution funds keeping, deposit at expiration, etc. Annual time deposit interest rate, for example, was 7 li (i.e. 7%), and its final accounting was settled every half year.

After its opening, the Joint Savings Society Bank's savings deposits had increased year by year. From 1923-1926, the total savings amount of Joint Savings Society

Side hall of the business hall

Bank had reached 90 million Yuan, which made it the biggest Chinese savings bank of the time. Apart form being placed with the four member banks, the banks' savings deposits were mainly utilized in making industrial and commercial mortgage loans, purchasing government bonds and other securities, as well as offering inter bank loans. The bank's business, however, had been gradually shrinking since the outbreak of the Pacific War in 1937.

The history of the Tianjin branch of The Northern Four Saving Banking Society includes an act of patriotism. Between 1932 to 1949, Mr. Chen Yihou, manager of Tianjin Branch of Salt Industrial Bank and Mr. Hu Zhongwen, manager of Tianjin Branch of The Northern Four Saving Banking Society hid the 16 gold chime bells, which served as collaterals when the last emperor Puyi asked for a loan from Salt Industrial Bank. The bells were kept in the storeroom of the basement of Tianjin Branch of Joint Savings Society Bank. There they successively avoided the search and robbery from the warlords, Japanese soldiers and police, and the KMT government. On January 18th, 1949, the third day after Tianjin's jiefang, the gold chime bells were finally taken out of hiding and presented to the state. Now the bells are on display in the Treasure Hall of the Forbidden City.

Built in 1923 and located at No.147 Jiefang North Road, It consists of a 3-story brick and concrete structure with a building area of 1,590 m². Its plan is in the shape of triangle. There is 1 arched door and 2 arched windows on the first floor, 4 Ionic columns on the second floor, and eaves on the top. The building is protected as a historical and stylistic architecture of Tianjin. It is currently occupied by the Tianjin Branch of the Industrial and Commercial Bank of China.

Original furniture

Meeting room

营业大厅柜台局部
A section of the counter in the business hall

营业大厅
Business hall

办公室楼道
Corridor

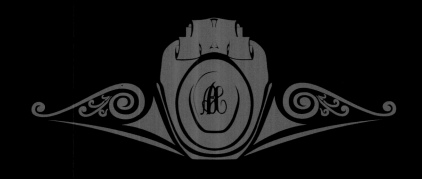

INTERNATIONAL BANKING CORPORATION, TIANJIN BRANCH

花旗银行

天津分行

昔日的花旗银行天津分行（今日的中国农业银行天津分行）
The former building of Citibank, Tianjin Branch (Now Agricultural Bank of China, Tianjin Branch)

营业大厅
Business hall

花 旗 银 行
天 津 分 行
INTERNATIONAL BANING CORPORATION, TIANJIN BRANCH

花旗银行创建于1901年，总行设在纽约。1916年花旗银行天津分行开业，当时注册资本300万美元。1941年太平洋战争爆发后被日军接收而关闭停业。1945年抗战胜利后，又在原址恢复营业，取代了英国汇丰银行而成为天津外商银行中的霸主。天津解放后，该行停业撤离。

花旗银行天津分行发行加印花旗银行天津地名的钞票有1元、5元、10元、50元、100元五种。1933年该行发行钞票达42.3万元。后来由于市民拒用，1934被迫陆续收回，停止流通。

花旗银行天津分行主要为美国商人对华贸易的金融周转业务提供多方面服务，在它的主要往来客户中绝大多数是经营中美贸易的进出口企业。1933年花旗银行天津分行资本、公积金和各项存款总计7 016.5万元。

花旗银行天津分行行址初设英租界中街"道济洋行"旧址，1921年迁入英租界中街66号新楼（今解放北路90号）。主楼三层，西洋古典复兴式建筑，设有地下室。门前有4根爱奥尼克柱，构成开放式柱廊，廊前铺筑欧式台阶。楼内有壁柱和顶部花饰。营业大厅内有7根方柱，内墙面有壁柱，顶部装饰华丽精巧。该建筑系天津市文物保护单位，特殊保护等级历史风貌建筑。现为中国农业银行天津分行。

花旗银行在天津发行的纸币
Banknotes issues by Citibank Tianjin Branch

上：大门门铃
Above: The door bell

下：通往办公室的楼梯
Below: Staircase leading to the offices

花旗银行外景
Citibank exterior

Headquartered in New York, the National Citibank (Citibank for short) was founded in 1901 with registered capital of $3 million. Citibank closed and halted its business because of being taken over by the Japanese Army after the outbreak of Pacific War in 1941, and resumed its business at the original site after the victory of Anti-Japanese

大厅一角
Grand hall

上：华丽的方柱
Above: Splendid square column

下：拱形的窗户
Below: Arched window

War in 1945. The bank surpassed the Hong Kong and Shanghai Banking Corporation to become the most powerful foreign bank in Tianjin, but ceased business and closed after Tianjin's jiefang.

Tianjin Branch of the National Citibank issued 5 versions of banknotes imprinted with Tianjin places names: 1 Yuan, 5 Yuan, 10 Yuan, 50 Yuan and 100 Yuan. Citibank issued banknotes amounting to 423,000 Yuan in 1933, but was compelled to withdraw those banknotes from circulation in 1934 since citizens refused to use them.

Tianjin Branch of the National Citibank mainly provided multiple services for American merchants trading with China. Its current accounts showed that most of the businesses were engaged in Sino-U.S. imports and exports trade. In 1933, Tianjin Branch of the National Citibank's equity funds, accumulation funds and deposits totaled 70.165 million Yuan.

Citibank initially located itself at the original site of "Dow Kee Foreign Firm" on Victoria Road in the British Settlement, and then in 1921 moved to a new building on No.66 Victoria Road (now No. 90 Jiefang North Road). Featuring classical style, the main body of the building has 3-stories and abasement. In front of the entrance stand 4 Ionic columns forming an open colonnade, in front of a flight of steps. Inside the building, there are pilasters, and flower decorations on the ceiling. In the business hall, there are 7 square columns, with pilasters in the inside walls. The building is protected as a Tianjin cultural relic. It is currently occupied by the Agricultural Bank of China, Tianjin Branch.

大楼外檐装饰
Bank exterior decoration

办公室走廊
Corridor

TUNG LAI BANK, TIANJIN BRANCH

东莱银行
天津分行

昔日东莱银行天津分行（今日的天津科学宫）
Formerly the Tianjin Branch of Tung Lai Bank (Now Tianjin Science Museum)

东莱银行
天津分行
TUNG LAI BANK, TIANJIN BRANCH

东莱银行创建于1918年2月1日，原始资本20万元，独资经营，总行设在青岛，济南、天津、上海、大连设有分行。1923年2月，该行改组为股份有限公司，增资到300万元。1926年2月该行总行移至天津，天津分行改为总行，行址在法租界21号路（今和平路383号）。1933年9月该行总行又移至上海。1952年12月该行参加全行业公私合营，与其他银行、钱庄一起组成统一的公私合营银行。

东莱银行主要办理存款、放款、汇兑、外币兑换和有价证券买卖业务。1932年该行存款余额954万元，贷款余额528万元，有价证券买卖128万元，领用兑换券25万元。

东莱银行1919年在天津设立分行，最初行址在宫北大街，1930年迁至和平路289号。该建筑由德国建筑师贝伦特(Behrent)设计，主楼为三层混合结构，局部四层，局部有地下室，占地面积3 214m²，建筑面积6 486m²，建筑形式为古典复兴形式，柱头为罗马混合式，采用中国方形回纹代替了圆形卷涡，故也可以称为折中主义建筑形式。主要入口在马路转角处，立面采用了古典主义的三段论设计。底层提高后形成台基，一、二层用混合式巨柱柱廊。第三层的檐部以上为阁楼层，并有方柱廊。在入口上部有三角形山花，阁楼层上部又有重檐的圆形塔楼，塔楼上面一层用圆柱子，圆柱内是透空的，重檐塔楼略似北京天坛的高坡圆顶，使入口处显得雄伟壮观，增加了建筑的层次感和立体效果。从入口门厅进入营业大厅，大厅的一边为接待室，另一边为经理室、接待室、保险库、办公室等，大厅中央设4根大理石柱，彩色水磨石地面，大理石护墙板，天花为石膏花饰。二层设大会议室、饭厅、总经理办公室等，三层是职工宿舍等。该建筑系天津市文物保护单位。现为天津银行。

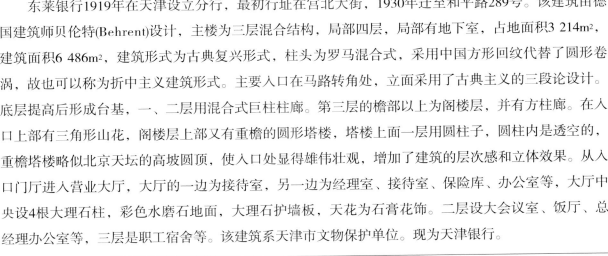

银行门前的柱廊
Colonnaded in front of the bank

上：西楼檐墙
Above: West building decoration

下：东莱银行外景
Below: Tung Lai Bank exterior

东莱银行营业大厅一角
Tung Lai Bank business hall

Tung Lai Bank was founded on February 1st, 1918 with original capital of 200,000 Yuan. Headquartered in Qingdao, the solely-funded bank set up branches in Jinan, Tianjin, Shanghai, and Dalian. In February 1923, the bank was restructured as a limited Company and increased its capital to 3 million Yuan. In February 1926, the bank moved its

银行门前的柱廊
Colonnaded in front of the bank

南楼入口
South building entrance

headquarters to Tianjin; located in Rue de Chaylard in the French Settlement (now No. 383 Heping Road). In September 1933, the bank moved its headquarters to Shanghai. In December 1952, the bank was nationalized by the communist government.

Tung Lai Bank's mainstream businesses covered deposits-accepting, loan-making, transfer, foreign currency converting and securities trading. In 1932, the bank balance of deposits was 9.54 million Yuan and the balance of loans 5.28 million Yuan, securities' trading volume 1.28 million Yuan and 250,000 Yuan for the exchange certificates.

a Chinese style pattern on the capitals. The main entrance is on a corner of the road. The ground floor is heightened to form the base and the huge columns span the first and the second floors. Above the entrance there are triangular shaped pediments and above the attic story there is a circular tower building with double eave roof, similar in style to the high-pitched circular roof of the Temple of Heaven in Beijing. Coming from the entrance doorway into business hall, on one side is the reception room and on the other side is the managers room, reception room, bank vault and offices. With four marble columns

南楼办公走廊
South building corridor

南楼与北楼造型各异的楼梯间
The stairwells of the south and north building, each has a very different style

In 1919, Tung Lai Bank set up its Tianjin branch, the site of which originally was in Gongbei Street, and then moved to No.289 Heping Road in 1930. Designed by Behrent, a German architect, the main building is of 3-stories with an auxiliary building of 4-stories and basement. The building covers an area of 3,214 m², and the floor space of 6,486 m2. The building is of classical style, the building adopts Roman columns with

standing in the middle, the hall is decorated with colorful terrazzo floors, marble wainscoting and gypsum flower ornament on the ceiling. On the second floor there is a conference room, dining hall and General Office, etc. The third floor is the employees' living quarters. There is a pedestrian bridge leading to the building. The building is classed as a Tianjin cultural relic. It is currently occupied by the Bank of Tianjin.

精美的楼梯扶手
Fine handrail

东楼的二楼过厅
Hallway on the east building second floor

南楼入口内景
South building entrance

会客室
Reception room

面积近百平方米的二楼会议室,装修设计仿造18世纪欧州宫廷模式,富丽华贵
One hundred square meter meeting room, in lavish eighteenth century style

CHINA-BELGIUM BANK, TIANJIN BRANCH

华比银行

昔日华比银行天津分行（今日的中国建设银行天津分行）
Formerly the Tianjin Branch of the China-Belgium Bank, Tianjin branch (Now China Construction Bank, Tianjin Branch)

华比银行
天津分行
CHINA-BELGIUM BANK, TIANJIN BRANCH

华比银行在天津发行的纸币
Banknotes issued by China-Belgium Bank, Tianjin branch

华比银行创办于1902年，资本为100万法郎。总行设在比利时首都布鲁塞尔，在我国上海、天津、北京、汉口、香港等地设有分行。华比银行天津分行于1906年开业，行址在英租界怡和道（今大连道），1922年迁至英租界中街86号（今解放北路104号）。该行于1941年太平洋战争爆发后停业，1945年抗战胜利后复业。1956年停业清理。

华比银行天津分行除经营存款、放款、汇兑等一般银行业务外，还从事投资铁路，承揽铁路借款等业务。比利时在天津设立的电车车灯公司及其他比商为该行主要客户。1949年1月15日天津解放后，该行曾被批准为经营外汇的"指定银行"，代理中国银行买卖外汇及国外汇兑业务。

华比银行天津分行发行的加印华比银行天津地名纸币银元券有1元、5元、10元、50元和100元共计10个版别。

华比银行大楼建于1922年，由天津义品房地产公司设计并监理，舜记营造厂承包施工。砖混结构，为带地下室的三层楼房，占地面积1 543m²，建筑面积3 663m²。该建筑外檐墙采用壁柱式花岗岩砌筑；正门为长方形，入口处砌扇形阶梯盘檐门楣；顶部为小檐平顶。建筑整体稳重简练，呈现代风格。该楼由华比银行和比利时领事馆共同使用，楼下为华比银行，楼上为比利时领事馆。该建筑系天津市重点保护等级历史风貌建筑。现为中国建设银行天津分行。

上：二楼阳台
Above: Second floor balcony

下：大楼侧面
Below: Bank exterior

营业大厅入口门厅
Entrance way of the business hall

China-Belgium Bank, Tianjin Branch was established in 1902 with its headquarters in Brussels. It had branches in Shanghai, Tianjin, Beijing, Hankou and Hong Kong. The registered capital was 1 million Franc. The Tianjin Branch of China-Belgium Bank opened in 1906 on the E-Wo Road in the British Settlement (now Da Lian Road), and moved to No.86 Victoria Road of British Settlement in 1922 (now No.104 Jiefang North Road). It was forced to close when the Pacific War broke out in 1941, and then resumed its business after the victory in the Anti-Japanese War in 1945. China-Belgium Bank liquidated and closed in 1956.

As well as normal commercial bank businesses such as deposit-accepting, loan-making and exchange, Tianjin branch of China-Belgium Bank also invested in the railway industry and specialized in railway loan-making. Its main customers were the

营业大厅
Business hall

Belgian light tram car company established in Tianjin and other Belgian merchants. After jiefang of Tianjin on January 15th 1949, Banoue Belge Pour L 'Etranger Bank was appointed as an authorized bank, surrogating for the Bank of China to trade foreign exchange in and out of the China.

The Tianjin branch of China-Belgium Bank issued silver dollar banknotes printed with Tianjin place names, namely one dollar, five dollar, ten dollar, fifty dollar and one hundred dollar.

Designed and supervised by Tianjin Credit Foncier D'Extreme Orient, the building of China-Belgium Bank was constructed by Shun Kee Construction Company in 1922. The main body is a 3-story building of brick and concrete structure. With an area of 1,543 m² and floor space of 3,663 m², the building's external wall has granite pilasters. In front of the rectangular entrance lays fan-like steps. With a flat top with small eaves, the building looks stable, modest and modern.

办公楼前厅
Entrance hall to the office building

前厅楼梯间
Stairwell in the entrance hall

The China-Belgium Bank and The Belgian consulate both occupied the same building. The first floor was for the China-Belgium Bank, and the second floor for the Belgian consulate. It is protected as an example of historical and stylistic architecture of Tianjin. The building is currently occupied by the China Construction Bank, Tianjin Branch.

会议室
Meeting room

昔日大陆银行天津分行（今日的中国交通银行天津分行）
Formerly the Tianjin Branch of the Continental Bank (Now China Bank of Communications, Tianjin Branch)

大陆银行
MAINLAND BANK

大陆银行由谈荔孙、张公权联合冯国璋、张勋等出资设立。1918年9月开始筹建，1919年4月开业，总行设在天津。资本初定200万元，实收100万元。1926年扩充为1 000万元，实收750万元。1930年增资收足1 000万股本。这个时期为大陆银行鼎盛时期，开办了许多附属机构。1935年大陆银行总经理处正式迁往上海。1952年12月该行参加全行业公私合营，与他它银行、钱庄一起组成统一的公私合营银行。

大陆银行开业时，大陆银行天津分行同时成立，并先后在北平、上海、汉口、南京、青岛、济南、蚌埠、杭州、南昌、苏州、郑州、长沙、哈尔滨、石家庄、太原、重庆等地设立分支机构。1931年天津分行改为管辖行，管辖哈尔滨、青岛、济南、蚌埠等行。该行自开办以来以中国银行为蓝本，各项业务以稳健发展为主旨。1919年大陆银行在京、津、沪三分行设立保管、信托专部，为市民保管贵重物品，开展买卖股票和有价证券等业务。1922年夏又于各分行内部设立储蓄专部。

大陆银行除办理一般商业银行业务外，兼办保管、信托、仓库事宜。1925年大陆银行存款余额2 038.9万元，放款余额1 399.5万元，纯收益为338.2万元。

大陆银行的仓库业务在各家银行中独具特色，1925年在天津建立大型仓库3处，租赁1处，大量存放商品货物。其栈租虽按同业中规定收费，但对本仓库出具的栈单作抵押借款时，利息加以优惠，以8扣办理。经过两年的努力经营，每年押款均达1 000万元以上。与此同时，大陆银行还在上海、汉口办理仓库业务，规模仅次于天津。

大陆银行旧址位于哈尔滨道68号，建于1921年。该建筑为砖混结构三层楼房，占地面积2 021m²，建筑面积4 895m²。方形门窗，水刷石外墙，欧式风格。该楼内装饰高级，设备完善。现门楣上"大陆银行"四字犹存。现为交通银行天津分行。

大楼上的方形塔楼
The square tower

上：办公室走廊
Above: Corridor

下：镶有彩色玻璃的圆拱形玻璃门
Below: Arch door with stained glass

The Mainland Bank was invested by Tan Lisun and Chang Kongchuan associated with Feng Kuo-chang and Chang Hsun. It was prepared and planned in September 1918, and opened in April 1919, with its headquarter in Tianjin. The initial planned equity fund was 2 million Yuan, and the actual paid-up capital was 1 million Yuan. In 1926, it was raised to 10 million Yuan with paid-up capital of 7.5 million Yuan. In 1930, the equity fund had been enlarged to 10 million Yuan. At this time, Mainland Bank was at its height of power and splendor, establishing many subsidiaries. In 1935, the General Manager Section of the Mainland Bank formally moved to Shanghai. In December 1952, the Bank was nationalized by the communist government.

After the Tianjin branch was established the Mainland Bank successively opened up branches in Peking, Shanghai, Hankou, Nanjin, Qingdao, Jinan, Bengfu, Hangzhou,

楼顶天窗
Sun roof

营业厅入口
Entrance to the business hall

营业大厅
Business hall

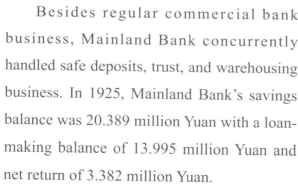

错层的经理室
Split-leveled General Managers Office

Nanchang, Suzhou, Zhengzhou, Changsha, Harbin, Shi Jiazhuang, Taiyuan, Chongqing, etc. In 1931, Tianjin Branch was upgraded to governing bank, charging branches of Harbin, Qingdao, Jinan, Bengfu, etc. From its establishment, Mainland Bank, modeling Bank of China, took modest and sound development as its main tenet. In 1919, it set up a special department of safe deposit and trust respectively in Beijing, Tianjin, Shanghai branches to keep valuables for citizens and develop business in trading stocks and other securities. In 1922, it set up another special department of savings in each branch.

Besides regular commercial bank business, Mainland Bank concurrently handled safe deposits, trust, and warehousing business. In 1925, Mainland Bank's savings balance was 20.389 million Yuan with a loan-making balance of 13.995 million Yuan and net return of 3.382 million Yuan.

Mainland Bank's warehousing business turned out unique among other Banks. In 1925, it established in Tianjin 3 big-sized warehouses and leased one to keep goods and commodities in large quantities. Although the rent was collected in accordance with the provisions within the industry, the interest would be preferential as 20% discount when using the indentures that the bank's warehouse made out as collateral to borrow money. Through 2 years' hard work, loans on security had reached more than 10 million Yuan per year. Meanwhile, Mainland Bank also handled warehousing business in Shanghai and Hankou.

Located at No.68 Harbin Road, the brick and concrete structured 3-story building was established in 1921 with a covering area of 2,021 m² and floor space of 4,895 m². It had square doors and windows, in a European style. The building had high-quality interior decoration and facilities. Even today, the four Chinese characters of the name of the Mainland Bank still exist above the door. The building is protected as a Tianjin cultural relic. It is currently occupied by the Bank of Communications, Tianjin Branch.

上：二楼楼道
Above: Second floor corridor

下：大理石楼梯
Below: Marble staircase

会议室
Meeting room

办公室楼梯间
The office staircase

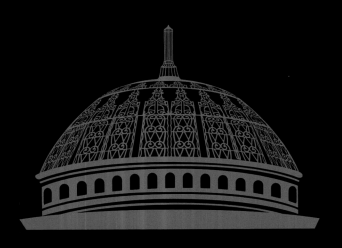

CHINA & SOUTH
SEA BANK, TIAN-
JIN BRANCH

中南银行

天津分行

昔日中南银行天津分行（今日的中国建设银行天津分行）
Formerly, Tianjin Branch of the China & South Sea Bank (Now China Construction Bank, Tianjin Branch)

中南银行 天津分行
CHINA & SOUTH SEA BANK, TIANJIN BRANCH

中南银行成立于1921年6月，总行设在上海。最大股东是1919年回国的印尼爪哇华侨黄奕柱与北京交通银行胡笔江创办的。中南银行的含义是中国与南海华侨合作之意，初建时银行资本额定为2 000万元，1924年实收750万元。1922年7月成立中南银行天津分行，行址在英租界中街（今解放北路98号），管辖北京支行。中南银行早期业务重心在天津、北京。1952年12月该行参加全行业公私合营，与其他银行、钱庄一起组成统一的公私合营银行。

中南银行经政府特许，享有兑换券发行权。为慎重纸币发行、兑换和保管准备金，1922年11月1日中南银行与盐业银行、金城银行、大陆银行联合成立天津四行准备库，规模宏大，专办保管准备金及发行事宜，地址在法租界21号路（今和平路、滨江道口）。天津四行准备库发行额至1935年时已达1 246万元。1935年币值改革时结束发行，天津四行准备库奉命关闭。1921~1931年期间，中南银行天津分行先后发行加印中南银行天津地名券1元、5元、10元、50元和100元共计11个版别。

中南银行天津分行下设储蓄部办理储蓄存款业务，放款部办理放款、投资业务，外汇部办理外汇业务，信托部办理信托业务。开业初期，该行存放款业务发展顺利。1925年全年决算津、京两地纯益之数较1924年溢出十分之六。1926年存款激增，其中定期存款年终余额达1 000万元。1936年存款余额为2 312万元，放款余额为1 395万元，全年损益-7.46万元。

中南银行天津分行大楼总体为古典主义风格，设计者受当时欧洲探新运动的影响，把古典主义建筑的部件进行简化，但基本上还受古典主义三段论的束缚。1938年由原来两层钢筋混凝土框架结构兼地下室一层，增建了第三层。在改造时因原来设计图纸找不到，由华信工程司沈理源进行测绘平面后设计加建。该建筑以半地下室作为台基。一、二层为柱身，半圆的柱子已完全简化，没有柱头，转角处的柱础成小八角板，柱墩亦呈小八角形。主要入口的二层顶部建有半球形的钢筋混凝土穹顶，穹顶外面套有一个铜质的镂花穹顶。受维也纳分离派的"整体简洁，集中装饰"的影响。从入口门廊进入大门厅，再经由多扇二道门进入营业大厅，中央用四根柱子支撑着三层大梁，并用栏杆围合，形成围廊。三层空间用墙与窗围合，从而形成底层地面直通三层玻璃穹顶的高大空间。营业大厅四周为经理室、接待室、会计室，二层回廊四周有会议室、办公室等，三层为职工宿舍等。该建筑系天津市文物保护单位，重点保护等级历史风貌建筑。现为中国建设银行天津市住房公积金管理中心。

上：顶部充满奥地利分离派风格的铸铜半球形穹顶
Above: Austrian style dome

下：三楼顶棚
Below: Third floor ceiling

银行大楼外景
Bank exterior

办公楼夜色
Night view of the office building

China & South Sea Bank Ltd. was founded in June, 1921 with its headquarter in Shanghai. The biggest shareholder was an overseas Chinese named Huang Yizhu who returned from Java Indonesia in 1919. The initial equity fund was 20 million Yuan, however, in 1924 the paid-up capital was only 7.5 million Yuan. In July 1922, Tianjin branch was founded at No.98 Victoria Road in the British Settlement (now Jiefang North Road). In the early days, the bank's business focus was in Tianjin and Beijing. In December 1952, the bank was nationalized by the communist government.

The China & South Sea Bank Ltd. through a government concession enjoyed the right of issuing exchange certificates. To take precaution on banknotes issuing, exchange and reserves keeping, on November 1st, 1922, the China & South Sea Bank Ltd. together

华丽的铁艺楼梯
Beautiful metal staircase

上：带护栏的窗户
Above: Window

下：铁艺楼梯扶手
Below: Handrail of the metal staircase

with the Salt Industrial Bank, Kinchen Bank and the Mainland Bank founded a joint reserve bank with abundant funds in Tianjin to specialize in keeping reserves and banknote-issuing business. Situated on Rue de Chaylard in the French Settlement (now the intersection of Heping Road and Binjiang Road), the joint reserve bank issued 12.46 million Yuan of banknotes up to 1935. The issuing ended after the currency reform in 1935, and the reserve bank closed as ordered. From 1921 to 1931, the Tianjin Branch of China & South Sea Ban Ltd. had successively issued banknotes printed with the name Tianjin, dominations of which were one Yuan, five Yuan, ten Yuan, fifty Yuan, one hundred Yuan, etc—11 versions in total.

Within the Tianjin Branch of the China & South Sea Bank Ltd. four departments were set up, namely, savings department to handle savings business, loan department to handle loan-making and investment business, foreign exchange department to handle foreign exchange business and trust department to handle trust business. Initially the bank's savings business went smoothly. In the 1925 year-end final accounting of revenue and expenditure, the net return of Tianjin and Beijing was 60% increase over 1924 figures. 1926 saw a dramatic increase in savings, where year-end time deposit balance reached 10 million Yuan. In 1936, savings balance was 23.12 million Yuan, loan-making balance was 13.95 million Yuan, and the whole year's profit and loss was -74,600 Yuan.

Affected by the European Search for New Architecture Movement, the building of the China & South Sea Bank simplified the classical architectural form. In 1938, Shen Liyuan of Hua Xin Engineering Company added the third floor to the previous 2-story reinforced concrete structure. Since the original blueprint had been lost, the third floor could only be added after surveying and drawing the plan. The architecture's semicircular columns were so simplified that there are no capitals on the top. The base of the columns is octagonal, as is column pier. An iron and concrete dome is situated above the main entrance. The dome has a copper coating engraved with Viennese style flower patterns. The entrance porch leads to an entrance hall which in turn through a folding door leads to the business hall. In the middle of the hall, the main beam is supported by 4 columns. On the second floor, there are another 4 columns supporting the main beam of the third floor. Between the columns, balusters are used to form a cloister. The hall has an impressive 3-story atrium from the ground floor direct to the glass roof of the third floor. Surrounding the business hall, there was the Managers Office, Reception Room, and Accounting Room; on the second floor, surrounding the cloister were Meeting Room, Offices, etc; while employees' living quarters were on the third floor. The building is protected as a Tianjin cultural relic. It is currently occupied by the China Construction Bank Housing Provident Fund Management Centre.

进入营业大厅的走道
Corridor to the business hall

二楼办公共享空间
Second floor open plan office

KOREAN BANK, TIANJIN BRANCH

朝鲜银行

天津分行

昔日朝鲜银行天津分行
The former Tianjin Branch of Korean Bank Corporation

朝鲜银行
天津分行
KOREAN BANK, TIANJIN BRANCH

朝鲜银行原为日本于1909年在汉城建立的韩国银行，1911年更名朝鲜银行，总行设在汉城。资本为1 000万日元。从1913年起陆续在中国上海、沈阳、大连、抚顺、长春、天津、北京、青岛、济南等地开设分行。1918年设立朝鲜银行天津分行，行址在法租界中街（今解放北路97号）。1945年抗战胜利后，朝鲜银行天津分行由国民政府指定中央银行天津分行接收。

朝鲜银行天津分行办理的主要业务有：存款、放款、贴现、汇兑和代理收付等。

朝鲜银行天津分行行址建于1918年。1938年迁至日租界旭街（今和平路）。该建筑为砖混结构，共三层，建筑面积2 700m²。该楼平面呈倒八字形，正巧门为菲律宾木雕花大门，两侧为14根陶立克式砖柱，整体为西洋古典风格，采用中式建筑的磨砖对缝传统手法，顶部出檐、坡顶。该建筑系天津市特殊保护等级历史风貌建筑。

二楼办公室的走廊
Second floor corridor

Korean Bank was originally the Republic of Korea Bank founded in Seoul by Japanese in 1909. It was renamed Korean Bank with its headquarter in Kyongsong in 1911. The equity fund was 10 million Japanese yen. From 1913 on, it had successively set up branches in Shanghai, Shenyang, Dalian, Fushun, Changchun, Tianjin, Beijing, Qingdao and Jinan, etc. In 1918, Tianjin branch of Korean Bank was set up on Rue de France of French Settlement (now No.97 Jiefang North Road). After the victory in the Anti-Japanese War in 1945, Tianjin branch of Korean Bank was taken over by Tianjin branch of the Central Bank of China under the instructions from the national government.

Tianjin Branch of Korean Bank's mainstream business covered as follows: deposits-accepting, loan-making, discounting, currency exchange and agency for receipts and payments, etc.

Located on Rue de France of French Settlement,(now No.97 Jiefang North Road), the building of this Japanese-funded bank was constructed in 1918, and moved to Asahi Road in the Japanese Settlement (now Heping Road) in 1938. It is a 3-story building of brick and concrete structure with a building area of 2,700 m². The building plan looks like a V-shape. The entrance is made of Philippine wood with flower carving, and 7 Doric brick columns stand along each side of the gate. Adopting the traditional method of rubbed bricks with tight joints in Chinese architecture, the entire building is of western classical style with top outstretched eaves and sloping roof. It is under the special-level protection as of historical, stylistic and architectural importance to Tianjin. It is current an office building.

方形的梁柱
Square beam

办公楼前厅
Front hall of the office building

INDO-CHINA AGENCY BANK, TIANJIN BRANCH

东方汇理银行

天津分行

东方汇理银行
天津分行
INDO-CHINA AGENCY BANK, TIANJIN BRANCH

法国东方汇理银行于1875年创办,总行设于巴黎。组织形式为股份有限公司。该行与中法实业银行为姐妹行,两行董事互有参股。清光绪三十三年(1907年)在津租用法租界西宾馆开设东方汇理银行天津分行,原始资金100万两白银。光绪三十四年(1908年)该行在法租界中街(今解放北路73号)购地建造大楼,1912年建成后迁入营业。东方汇理银行天津分行曾代表法国与英、美、德、日组成五国银行团,1913年借予袁世凯2 500万英镑的"善后大借款",并继续营业。1945年抗战胜利后,该行业务发展迅速,在津取得外商银行的首席地位。1956年该行宣告停业,是天津解放初期最后关闭的一家外商银行。东方汇理银行天津分行的主要业务范围是存款、放款、汇兑、贴现、经营进口、出口押汇和买卖外汇。

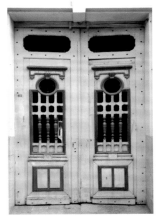

后侧门
Rear side door

东方汇理银行天津分行1908~1921年由比利时义品公司设计,设计人是查理(Charrey)及康沃西(Conversy),由义品公司工程部监督施工。该建筑为三层砖木结构,带半地下室。占地面积1 244m²,建筑面积3 651m²,为古典折衷主义建筑形式。古典三段论很明确,半地下室做台基。仿石块砌筑,深横竖缝,低层墙面用仿石块砌筑,深横缝大厅地面铺彩色地砖。圆拱形窗,窗外有精致铁花。墙面用红砖砌,窗户有瓶饰栏杆,二层窗户下用瓶饰,窗户两侧用长牛腿支撑着三层的阳台。三层有带瓶饰阳台,檐部出檐很小,檐部上的女儿墙带有瓶饰栏杆。原建筑在沿马路屋顶角上有3个亭子,因1976年震损拆除。该建筑是天津市重点保护等级历史风貌建筑。

银行大厅正门
The front door of the grand hall of the bank

银行大楼外景
Bank exterior

Indo-China Agency Bank (Calyon Bank for short) was founded as a limited company in 1875, with its headquarters in Paris. As sister banks, directors of Calyon bank and China-France Industrial Bank held each other shares. Calyon bank came to Tianjin in 1907, renting the West Hotel in the French Settlement to open Indo-china Agency

Second floor corridor

Fine handrail

Bank, Tianjin Branch. The original capital was 1 million silver liang. In 1908 the bank bought a plot of land at Rue de France in the French Settlement (New No.73 Jiefang North Road) to construct the building. In 1912 the building was completed and the Bank moved in. Indo-china Agency Bank, Tianjin Branch, on behalf of France, associated with Great Britain, USA, Germany, Japan to form a five-nation banking consortium, which made the "Big Loan" of 25 million pound to the Yuan Shikai Administration in 1913. During the period when Tianjin was under Japanese occupation, Calyon bank fortunately avoided being taken over by Japanese, and so was able to keep continue business. After the victory in the anti-Japanese war in 1945, the bank's businesses developed rapidly, winning the first place among foreign banks in Tianjin. Calyon bank ceased business in 1956, the last foreign bank to close after Tianjin's jiefang. The mainstream businesses of the Indo-china Agency Bank, Tianjin Branch covered: deposit-accepting, loan-making, remittance, discounting, inward/outward document billing, and foreign exchange trade.

From 1908 to 1921, the Belgian Company Credit Foncier D'Extreme Orient had only completed a part of the building, according to the design paper provided by French general headquarters of Banque Del' Indochine in Paris. It was designed by Charrey and Conversy, supervised and constructed by engineering department of Credit Foncier D'Extreme Orient. The building is a 3-story brick-timber structure with a covering area of 1,244 m² and a floor space of 3,651 m². The architecture is in the form of eclecticism. The building was built with stone with deep horizontal and vertical joints. The windows are circular arched, framed with a delicate ironwork flower design. The second and third floors are constructed from red brick. Originally, there were 3 pavilions at the corner of the roof along the road, which were torn down due to the earthquake in 1976. The hall's floor is laid with colorful tiles. The building is protected as a historical and stylistic architecture of Tianjin.

Business hall back door

营业大厅
Business hall

楼梯间
Staircase

办公室楼梯间
The office staircase

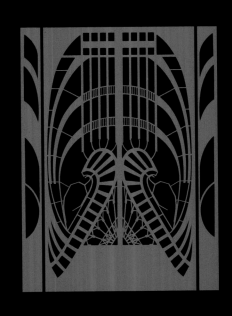

XIN HUA TRUST & SAVINGS BANK, TIANJIN BRANCH

新华信托储蓄银行

天津分行

昔日的新华信托储蓄银行天津分行
The former Tianjin Branch of the Tianjin Branch Of Xin Hua Trust & Saving Bank

新华信托储蓄银行
天津分行
XIN HUA TRUST & SAVINGS BANK, TIANJIN BRANCH

新华信托储蓄银行，原名新华储蓄银行，由中国、交通两行合拨资金100万元，于1914年10月20日创办，总行设于北京。1925年改名为新华商业储蓄银行。1931年改组为新华信托储蓄银行。总行迁设上海，由官办转为商办。1948年11月改名为新华信托储蓄商业银行，为当时的"南四行"（浙江兴业银行、浙江实业银行、上海商业储蓄银行及新华信托储蓄银行）之一。1917年设立新华信托储蓄银行天津分行，行址在法租界中街（今解放北路17号）。1936年初迁入法租界中街新华大楼（今解放北路14号）。天津分行下设敦桥道办事处、梨栈办事处、旭街（今和平路）办事处和河北大街办事处。

新华信托储蓄银行天津分行内部设信托、储蓄两部。信托部主要办理定期存款、活期存款、各项放款、汇兑、代理买卖各种证券及中国公司股票、经营房地产和仓库业以及代理收付款等；储蓄部主要办理生活储金、人寿储金、教育储金、存取两便储金、零存整取储金、整存整取储金、整存零取储金及存本付息储金等。1921年该行发行流通储蓄金券，在京、津、沪三地流通甚广。该行储蓄种类繁多，手续简单方便，服务热情周到，深受客户欢迎，1938年存款额在同业中（除中央、中国、交通、农民四行外）居首位。

该建筑于1934年由华信工程司沈理源设计，1935年完成。受当时欧洲新建筑运动思潮影响，沈理源在设计新华信托储蓄银行时力求摆脱古典主义的影响，设计形式趋于现代建筑风格。建筑立面简洁，主体建筑六层，局部八层为退层塔楼，地下室一层，钢筋混凝框架结构，建筑面积7 026m²。底层外墙面用浅红色花岗岩砌筑，梁上用石雕花饰分隔，犹如古典建筑的台基。二至六层形成一体，用大块仿石面饰面，分成跨向，用巨柱式壁柱分隔，略微突出墙面，直上女儿墙。跨间两窗之间有贯通的细长壁柱，整个建筑采用竖线条，有向上冲的感觉，增加了建筑物的高度感，六层窗上的墙面有小方花饰。从二层到六层上下窗间铸铁图案铜饰面板十分醒目金光闪闪，表现出银行的雄厚财力。大门设在转角处，铜门花饰受新艺术运动的影响。营业大厅的八根方柱及地面、墙壁均用大理石铺砌，营业大厅面积的200m²，大厅四周布置经理室、文书室、会计室、会客室等。四层到五层为职工俱乐部和礼堂等由滨江道侧门入口到二层为职工宿舍三到六层为办公室，地下室为保险库、锅炉房等。该建筑系天津市文物保护单位，为天津市重点保护历史风貌建筑。

俱乐部大厅壁灯
The wall lamp of the club hall

楼顶上的双层塔楼
The double deck tower on the top of the building

Xin Hua Trust & Savings Bank Ltd. whose original name was Xin Hua Savings Bank, was founded on October 20th, 1914 by the Bank of China and the Bank of Communications. Both of which jointly channeled funds of 1 million Yuan. The headquarters was set up in Beijing. In 1925, its name was altered into Xin Hua Commercial Savings Bank. In 1931, it was restructured into Xin Hua Trust & Savings Bank Ltd. Its headquarters, thus, moved into Shanghai, and there was a shift from government-run to private-run. In November, 1948, its name was changed into Xin Hua Trust & Saving Commercial Bank, which was one of "the Southern Four" (National Commercial Bank, as well as Zhejiang Industrial Bank, Shanghai Commercial Savings Bank, and Xin Hua Trust & Savings Bank Ltd). In 1917, Tianjin branch was set up, located at Rue de France in the French Settlement (now No.17 Jiefang North Road). At the beginning of 1936, it moved into Xin Hua

东侧大门
East gate

办公室一角
Office

Building (now No.14 Jiefang North Road) on Rue de France in the French Settlement. Under the Tianjin branch, it set up Tunbridge Road (now Xi'an Road) Office, Lizhan Office, Asahi Road (now Heping Road) Office and Hebei Dajie Office.

Within Xin Hua Trust & Savings Bank Ltd. trust and savings departments were set up, mainly to handle time deposits, demand deposits, loan-makings, exchanges, acting trading securities as well as Chinese companies' stocks, real estate, warehousing, acting collection and payment, living deposits, life insurance deposits, educational deposits, saving/withdrawal deposits and small savings for lump-sum withdrawal deposits. In 1921, the bank issued and circulated savings golden banknotes, which were very popular in Beijing, Tianjin, Shanghai. The bank had many types of savings with simple and convenient procedures as well as a high level of customer friendly service. In 1938, its savings amount ranked among the top five banks.

Xin Hua Trust & Savings Bank Ltd. Building was designed by Shen Liyuan of Hua Xin Engineering Company and completed in 1935. Influenced by the European Modernist Movement, Shen Liyuan design was of a modern architecture style. The main body was a 6-story building with an 8-story tower at the rear. The building is constructed of reinforced concrete with a floor space of 7,026 m2. The ground floor is of light red granites with stone flower ornaments almost in classical style. The second to sixth floors are unified with a large imitation stone facade with huge columns and pilasters that project beyond the wall and directly up to the parapet. The whole design adopts vertical lines with the feeling of reaching up to the sky, creating a sense of height. There are small square flower ornaments on the wall above the windows of the sixth floor. The glittering copper panels with cast-iron patterns between the windows of second to the sixth floor are striking, showing off the bank's financial strength. The door is on a corner and has copper flower door ornaments influenced by Art Nouveau. The business hall covers an area of 200 m2, the floor and the wall were laid with marble, and there are 8 pillars. Off the business hall are the manager's office, secretary room, accounting room and reception room, etc. From the fourth to the fifth floor, there is employees' club and auditorium. The employees' living quarters are from the side door on Binjiang Road to the second floor. The third floor to the sixth floor is offices. The basement includes the bank vault and boiler room, etc. The building is protected as a Tianjin cultural relic.

大楼外墙花饰
Exterior decoration

上：四楼带电梯的过道
Above: Fourth floor elevator lobby

下：俱乐部二楼回廊
Below: Second floor corridor

东侧二道门
East side door

带有回廊的俱乐部大厅
The club hall corridor

原汇记钱庄
Formerly Huiji Money shop

原义品放款银行天津分行
Formerly Yipinfangkuan Bank, Tianjin Branch

解放路（原英租界中街）
Jiefang road (former Victor Road)

解放路（原英租界中街）
Jiefang road (former Victor Road)

原大通银行天津分行
Formerly Datong Bank, Tianjin Branch

原广瑞银号
Formerly Guangrui Money shop

杨福荫路钱庄街
money shops street on Yangfuyin road

夜幕降临,解放路金融街上仍然是金光灿灿、车水马龙,昔日"东方华尔街"将迎来新的辉煌

When the dark falls, Jiefang Road remains its bustling still; the former "East Wall Street" is welcoming the new flourishing tomorrow.

原中国农工银行天津分行
Formerly the Agricul Tural and Industrial Bank of China, Tianjin Branch

原中法储蓄会天津分会
Formerly China-France Saving Banking Society, Tianjin Branch

解放路（原法租界中街）
Jiefang road (former Rue de France)

原中国实业银行
Formerly the National Industrial Bank of China

原聚成兴银行天津分行
Formerly Juchengxing Bank Tianjin Branch

原中国实业银行
Formerly the National Industrial Bank of China

附 录
APPENDIX

近代天津银行一览表
A TABLE OF ALL THE MODERN BANKS IN TIANJIN

银行名称	津行成立日期	津行停业日期	最初资本额	首任负责人	津行经理	津行地址	总行所在地	总行成立日期	备注
中国通商银行	光绪24年（1898年）1947年4月复业	1905年撤销 1952年12月15日合并	额定500万两	天津分董 冯商盘	洋大班 厚士敦 华大班 梁景和	东浮桥北沿河马路大摆渡口附近（今狮子林桥西）复业后行址为中正路44号（今解放北路38号）	上海	1897年5月	享有纸币发行权
北洋天津银号（天津官银号）	光绪28年8月（1902年）	1901年9月改组为直隶省银行	官本50万两	督办 周学熙		东北角三义庙（北马路东端南侧）	天津	1902年8月《天津通志·大事记》为1903年4月1日	享有发行权
志成银行	光绪29年8月（1903年）	1911年10月通志大事记（1915年）	官股20万两 商股15万两	总董 杨俊元	张作涛	宫北大狮子胡同	天津	1903年8月	
户部银行	光绪31年10月（1905年）	1908年改为大清银行	额定400万两 实收200万两	总办 张允言		估衣街，1906年迁至北马路	北京	1905年	享有发行权
交通银行	光绪34年3月14日（1908年）	1912年改为中国银行	500万两	总理 李经楚		北马路，后迁法租界5号路与6号路拐角，继迁至法租界4号路（今滨江道）48号	北京 1928年迁上海	1908年	享有发行权
厚德商业银行	宣统元年5月（1909年）	1949年1月 1911年	100万两	总经理 王德基		针市街	北京	1909年5月	
信义银行	宣统元年（1909年）	1909年6月	10万元				江苏镇江	1906年11月	通用票 挤兑倒闭
公益商业储蓄银行	宣统元年12月（1910年1月）	1920年前	100万两	总理 李禔		北门内大街	北京	1908年7月	
北洋保商银行	宣统二年（1910年）3月1日	1939年1月	400万两		华经理 叶兰舫 外经理 瑞亨	北马路1926年迁法租界14号路（今承德道），1931年迁法租界中街52号（今解放北路）	天津 1921年7月迁北京	宣统二年（1910年3月1日）。1920年7月由中外合资改为华资银行	有纸币发行权
信成银行	宣统二年（1910年）5月	1912年2月	50万元	总理 周廷弼(舜卿)	陈溢庆	宫北大街	上海	1906年4月	有发行权
直隶省银行	宣统二年（1910年）10月	1928年	200万两（一说190万元）		徐树恒	北马路，（由天津官银号改组为直隶省银行）	天津	1910年10月	享有发行权
新茂银行	（大约）宣统年间设立	（约）辛亥革命前后歇业					天津		《北京金融志》96页
殖业银行	1911年3月19日 1946年8月复业	1945年5月停业 1947年6月清理	额定200万元 实收108.1万元	总理 李士钰		北门内只家胡同，1927年前后迁至法租界4号路100号（今滨江道）	天津	1911年	
兆丰银行	1912年6月	1920年前	3万元			北门外针市街			
中国银行	1912年10月2日	1949年1月	6 000万元	行长 唐瑞铜	林葆恒	由大清银行改组成立，行址在法租界7号路（今解放北路）24号，1918年4月迁法租界8号路（今赤峰道）	北京 1928年迁上海	1912年8月	有货币发行权
殖边银行	1914年11月《天津通志大事记》为1915年8月11日	1925年	额定2 000万元 实收200万元			宫北大街	北京	1914年11月	有纸币发行权
盐业银行	1915年5月29日	1952年12月15日	500万元 先收四分之一	总经理 张镇芳	张作涛 (松泉)	法租界8号路（今赤峰道12号）	北京，1928年8月迁天津，1934年6月迁上海	1915年3月	
浙江兴业银行	1915年10月24日	1952年12月15日	额定100万元 先收四分之一		潘履园	初设宫北大街路西354号，1925年迁至法租界21号路与26号路转角（今和平路237号增1号，原为319号滨江道口）	杭州1915年后移上海	1907年5月27日	有纸币发行权
山西裕华银行	1915年11月 1947年1月复业	1942年6月停业 1949年1月被接收	额定500万元			初在宫北大街，后迁法租界8号路46号（今赤峰道）	山西太谷，1927年天津，1946年迁上海	1915年	
蔚丰商业银行	1916年5月18日	1924年9月	额定300万元		梁纬堂	日租界旭街（今和平路）	北京	1916年由蔚丰厚票庄改组成立	
财政部 平市官钱局	1916年6月10日			监督 陈福颐		鼓楼大街路东67号	保定	1914年11月	发行铜元票纸币
中孚银行	1916年11月7日	1952年12月15日	额定200万元 先收四分之一	董事长 兼总经理 孙荫亭(多森)	沈伯循 (继武)	北马路，1918年10月迁法租界8号路27号（今赤峰道）	天津 1930年迁上海	1916年11月7日	1925年获准发行纸币权
金城银行	1917年5月15日	1952年12月15日	额定200万元 先收四分之一	总董 王郅隆(祝三) 总经理 周作民	阮寿岩 (福镛)	法租界七号路43号，1921年2月迁英租界中街20号（今解放北路108号）	天津 1936年迁上海	1917年5月15日	
中兴银行						1917年2月21日天津县银钱业未注册行号的清单所列			

近代天津银行一览表
A TABLE OF ALL THE MODERN BANKS IN TIANJIN

银行名称	津行成立日期	津行停业日期	最初资本额	首任负责人	津行经理	津行地址	总行所在地	总行成立日期	备注
义兴银行		1929年8月2日				1917年2月21日天津县银钱业未注册行号的清单所列	北京		
新华信托储蓄银行	1917年7月23日	1952年12月15日	额定100万元 先收15万元	总理 方仁元	马殿元（文卿）	法租界7号路17号(今解放北路)，1936年初迁入法租界中街新华大楼(今解放北路14号)	北京 1931年迁上海	1914年10月20日	
山东银行	1918年	1925年改为山东商业银行	额定500万元 已收141.12万元	总理 张子衡		宫北大狮子胡同	济南		有代理金库发行纸币权
山东工商银行	1918年	1929年	额定200万元 先收52万元			北马路	济南	1918年	
聚兴诚银行	1918年6月 1946年8月复业	1930年撤消 1952年12月15日	额定100万元	主席 杨文光 总经理 杨希仲	杨芷芬	法租界中街19号(今解放北路)复业时在1区兴安路277号	重庆	(1914年创立) 1915年3月16日	
中法储蓄会	1918年8月	1935年7月由中央信托局接办	20万元			法租界马家口1号路(今大沽路)	北京	1918年8月	
五族商业银行	1918年9月	1928年	额定100万元 先收42万元	总理 陈文泉		估衣街	北京	1918年9月6日	
东莱银行	1919年3月1日	1952年12月15日	20万元	刘子山	高樾栽	宫北信成里，1921年迁宫北大狮子胡同，1925年迁法租界21号路(今和平路383号)	青岛 1926年迁天津 1933年迁上海	1918年2月	
大生银行	1913年3月8日	1949年1月	额定200万元 先收60万元	总理 苏国华	魏长源	北马路，1930年后迁法租界6号路61号(今哈尔滨道33号)	天津	1919年3月8日	
大陆银行	1919年4月1日	1952年12月15日	额定200万元 先收100万元	首席董事兼总经理 曹国嘉	总行经理 曹国嘉	法租界6号路(今哈尔滨道70号)	天津 1940年迁上海	1919年4月1日	
中国实业银行	1919年4月26日	1952年12月15日	200万元	总董 熊希龄 总理 周学熙	巢凤冈（季仙）	法租界12号路26号(今营口道)1923年4月7日迁法租界领事道该行新楼(今大同道13号)	天津 1932年4月迁上海	1919年4月26日	有发行纸币权
齐鲁银行	1919年8月前		额定100万元	经理 朱桂山		锅店街	济南	1919年	
边业银行	1919年8月创立 1920年4月改组	1937年10月	额定1 000万元 实收255.9万元	总理 张琛	杨承甫	法租界巴黎路88号(今吉林路) 1936年迁法租界8号路117号(今赤峰道)	北京，1925年改天津，1926年移沈阳，1931年迁回天津，1936年迁上海	1919年	有发行纸币权
北京商业银行	1919年 1946年9月复业	1927年5月停业 1949年1月	额定100万元 实收50万元	总理 张肇达	张昭文	法租界六号路(今哈尔滨道) 1921年12月迁法租界5号路与14号路转角(今吉林路与承德道口) 复业后在哈尔滨道60号	北京	1918年12月9日	
大中银行	1919年 (一说津行设立于1929年6月)	1950年3月27日	150万元	总理 汪云松		法租界马家口1号路(今大沽路)，1935年迁法租界中街15号(今解放北路52号)	重庆 1929年迁天津 1934年迁上海	1919年	有发行纸币权
国民银行	1919年		30万元	总理 许世英	金猷澍	意租界	天津	1919年	
明华银行	1920年6月	1935年5月	额定200万元 已收100万元	董事长 童金辉 总经理 童金吾	童梦熊	法租界4号路26号(今滨江道)	北京 1925年迁上海	1920年6月	
丰业银行	1920年9月 (办事处)	1938年	额定100万元 已收20万元	董事长 蔡成勋		针市街	归绥	1920年	有发行纸币权
上海银行	1920年11月	1952年12月15日	10万元	总董 庄得之 总经理 陈光甫	杨子和	宫北大街狮子胡同，1924年将办事处改分行迁东马路，1926年迁法租界8号路100号(今赤峰道36号)	上海	1915年6月	
东三省银行	1920年12月3日	1924年7月合并于东三省官银号	额定800万元 已收438万元	督办 张之汉 总办 刘尚清	孙云清	法租界5号路(今吉林路)	哈尔滨 1924年6月移长春	1920年10月	有发行纸币权
致中银行		1923年8月29日	100万元	董事长 赵椿年			北京	1920年	
东陆银行	1921年3月15日	1925年2月	额定200万元 已收100万元	总理 贺得霖	李健如	海大道8号(今大沽路)，1923年迁英租界中街该行新楼(今解放北路147号)	北京 1924年迁天津	1919年5月	
中华储蓄银行	1921年4月19日	1924年6月	额定100万元 已收34万元	总董 刘文揆	王国铎	英租界中街，法租界花园街，后迁锅店街	北京	1919年 (1925年停业清理)	
劝业银行	1921年5月27日	1931年	额定500万元 已收239万元	董事长 潘复		英租界海大道(今大沽路)，1922年10月迁法租界8号路32号(今赤峰道)，后迁法租界12号路戒酒楼(今营口道)	北京	1920年10月	有发行纸币权
大业银行	1921年7月11日	1928年7月	100万元			东马路	天津	1921年	
农商银行	1921年7月16日	1929年3月	额定1 000万元 已收170万元	总裁 齐耀珊		法租界6号路(今哈尔滨道)，1923年11月迁法租界中街(今解放北路与赤峰路口)	北京	1921年7月	有发行纸币权

近代天津银行一览表
A TABLE OF ALL THE MODERN BANKS IN TIANJIN

银行名称	津行成立日期	津行停业日期	最初资本额	首任负责人	津行经理	津行地址	总行所在地	总行成立日期	备注
裕津银行	1921年8月13日	1951年10月	额定100万元 先收30万元	总董 魏信臣	沈雨香	宫北大街，1941年后迁法租界8号路32号(今赤峰道)	天津	1921年7月	
裕丰银行						针市街			
同利银行						法租界			
同义银行						英租界			
天津工商银行						日租界			
华胜银行						宫南			
华充银行						宫北			
裕华银行						意租界			
通昌银行						宫北新街			
华昌银行						法租界中街			
天津通易银行	1921年10月	1922年12月	额定300万元 先收四分之一	总理 张澹如	顾仲甫	宫北新街	天津，1921年11月改为上海	1921年10月	
天津兴业银行	1921年10月	1928年	额定200万元 先收四分之一		李华亭	法租界	天津	1921年10月	
华新银行	1921年	1931年	100万元	总经理 叶之樵	朱幼樵	法租界6号路(今哈尔滨道)，后迁承德道	天津	1921年	
天津华北银行	1921年11月9日	1929年	100万元	总董 张弧 总经理 章勋	韩文涛	宫北	北京	1921年	华北银号改组成立
华孚银行	1921年	1922年	额定100万元 已收45.85万元	董事长 盛泽承		法租界海大道(今大沽路)	上海(初时在杭州)	1917年	
宝通银行	1921年		额定100万元 先收四分之一				天津	1921年	
裕达银行	1921年	1934年前已停业					天津	1921年	
直隶省官钱局	1921年11月11日	1935年	100万元	总理 张济川		宫北	天津	1921年	发行纸辅币
中南银行	1922年7月5日	1952年12月15日	额定2 000万元 先收500万元	董事长 黄奕住 总经理 胡笔江	王孟钟	英租界中街98号(今解放北路88号)	上海	1921年6月	有发行纸币权
北京裕华银行	1922年10月(分庄)	1923年秋	额定100万元 实收20万元	总理 魏福欧		日租界大和街(今海拉尔道)	北京(1923年秋停业)	1922年4月19日	
山西省银行	1922年10月	1937年	额定300万元 已收117万元	总理 阎维藩		针市街	太原	1919年	有发行纸币权
四行准备库	1922年11月	1935年			张泽湘(召兰)	法租界21号路63号(今和平路)1925年迁英租界中街67号(今解放北路147号)	北京后迁上海	1922年9月	发行中南银行纸币
察哈尔兴业银行	1922年(原汇兑所改分行)	1924年	100万元			锅店街，后迁东门外单街子	张家口	1916年(1924年休闭后，1925年10月1日复业)	有发行纸币权
天津道生银行	1922年12月19日	1928年10月	120万元			北马路	天津	1922年12月	
中华统一银行	1923年						天津		
四行储蓄会	1923年6月	1948年	100万元	主任 吴鼎昌	张召兰	法租界21号路63号(今和平路)1925年迁英租界中街67号(今解放北路147号)	上海	1923年6月	
蒙藏银行	1923年6月	1929年	额定1 000万元 已收500万元	总理 陈廷杰		法租界12号路(今营口道)	北京	1923年6月	有发行纸币权
怀远银行	1923年	1925年6月	额定500万元 已收125万元	总理 杨韶九	严柳村	法租界8号路(今赤峰道)	天津	1923年	
热河兴业银行	1923年	1928年查封 1931年停业	额定100万元			宫北新街 后迁法租界12号路(今营口道)	承德	1919年	有发行纸币权
普益银行	1923年					日租界旭街(今和平路)	天津	1923年筹设	
民国银行	1923年		额定500万元 先收200万元				北京	1923年	
大有银行	1924年2月	1927年4月	额定100万元 已收50万元			宫北大狮子胡同	北京(章程原定天津)	1923年	
甘肃省银行	1924年3月1日		额定100万元 已拨14.2万元			宫北大街	兰州	1923年	有发行纸币权
河南省银行	1924年3月	1927年	额定500万元 已收125万元			宫北大街	开封	1923年7月	有发行纸币权

近代天津银行一览表
A TABLE OF ALL THE MODERN BANKS IN TIANJIN

银行名称	津行成立日期	津行停业日期	最初资本额	首任负责人	津行经理	津行地址	总行所在地	总行成立日期	备注
香港工商银行	1924年	1930年7月	额定500万港洋 已收116万港洋			法租界海大道44号(今大沽路)	香港	1917年	
直东地方实业银行	1924年6月筹设						天津	1924年6月筹设	
大成银行	1924年		额定100万元			北马路,后迁法租界	北京	1921年	
民利银行	1924年					日租界			
东方商业银行	1925年5月4日	1926年6月自行清理1928年5月停业	额定500万元港元 已收162万元			法租界6号路82号(今哈尔滨道)	香港	1922年	
香港国民商业储蓄银行	1925年6月 1936年3月复业	1935年9月16日停业	额定200万元港币 已收200万元			法租界26号路45号(今滨江道),1935年迁法租界8号路103号(今赤峰道)	香港	1921年12月	
中元实业银行	1925年	1927年5月	额定400万元 先收四分之一			大沽路	天津	1925年9月	有发行纸币权
奉天商业银行	1925年7月						沈阳		
天津农工银行	1925年9月筹设	1926年2月退股解散				(因时局不靖,开办无期,乃停办退股)	天津	1925年9月筹设	
华威银行	1925年10月15日	1928年12月	额定1 000万元 已收250万元		郭子谨	法租界4号路(今滨江道)	北京	1922年	有发行纸币权
山东商业银行	1925年 由山东银行改称	1930年3月	额定500万	总理 张肇铨		宫北大狮子胡同	济南	1925年 由山东银行改名	
太平银行	1925年	1925年	额定500万元 先收四分之一				北京	1925年	
漳夏实业银行	1925年								
中国丝茶银行	1926年1月	1928年5月	额定500万元 已收125万元	总裁 陈金鼎		法租界24号路(今长春道)杨福荫里	天津	1926年1月	有发行纸币权
西北银行	1926年2月22日	1926年7月1日奉命停业			贺苻舫	法租界21号路(今和平路)	张家口	1925年4月	有发行纸币权
中国垦业银行	1926年3月15日	1952年12月15日	额定500万元 先收125万元	董事长 童金辉 总经理 俞佐庭	竺玉成	法租界8号路(今赤峰道)1927年迁6号路82号(今哈尔滨道34号)	天津 1929年改组迁上海	1926年3月	有发行纸币权
山东省银行	1926年8月	1928年4月				法租界	济南	1925年9月	有发行纸币权
中国农工银行	1927年2月	1951年2月14日	额定500万元 先收四分之一			法租界14号路24号(今承德道)后迁法租界中街(今解放北路63号)	北京 1929年迁天津 1931年移上海	1927年2月	有发行纸币权
东亚商业银行	1927年		100万元				北京	1927年10月	有发行纸币权
中兴银行	1927年10月						天津	1927年10月	
河南省农工银行	1929年2月 (办事处)	1936年9月裁撤	500万元			日租界乾泰栈内,后迁英租界2号路信义里6号(今营口道),1934年12月迁至英租界广东路福荫里8号	开封	1928年3月	发行辅币券
天津商业银行	1929年		额定100万元 先收50万元			英租界西摩路前协和贸易公司旧址	天津	1929年	
河北省银行	1930年1月	1949年1月接管	额定400万元 实收144.86万元	行长 梁新明		东北城角,1931年11月迁英租界11号路福善里1号,1932年5月迁法租界8号路66号(今赤峰道),1936年迁法租界14号路(今承德道),1946年迁中正路74号(今解放北路)	北京 1930年移天津	1929年3月,1930年冬改组改称	有发行纸币权
中央银行	1931年4月10日 1945年11月6日复业	1937年裁撤 1949年1月接管	2 000万元	总裁 宋子文	李达 (宏章)	英租界中街9号(今解放北路117号)	上海	1928年11月1日	发行法币
久安信托公司 (久安商业银行)	1931年6月	1952年12月15日	8万元	董事长 周止庵	尤士琦	法租界14号路34号(今承德道),1933年11月迁法租界英国大院12号,1942年迁英租界中街(今解放北路94号)	天津	1931年6月	1943年1月改称久安商业银行
陕西省银行	1931年8月 (办事处)		额定500万元 先收125万元	董事长兼总经理 韩光琦		法租界24号路(今滨江道),1935年迁法租界37号路安养里2号(今南京路),1937年迁法租界海大道(今大沽路)	西安	1930年12月	
中国国货银行	1931年9月	1943年1月	额定2 000万元 实收500万元	董事长 孔祥熙	温襄忱	法租界8号路110号(今赤峰道38号)	上海	1929年11月15日	
中原商业储蓄银行	1931年11月2日	1949年1月	额定100万元 实收50万元			日租界福岛街(今多伦道),后改1区中正路54号(今解放北路),(津分行曾在法租界26号路即滨江道)	天津	1931年11月	
银行名称	津行成立日期	津行停业日期	最初资本额	首任负责人	津行经理	津行地址	总行所在地	总行成立日期	备注

近代天津银行一览表
A TABLE OF ALL THE MODERN BANKS IN TIANJIN

银行名称	津行成立日期	津行停业日期	最初资本额	首任负责人	津行经理	津行地址	总行所在地	总行成立日期	备注
河北民生银行	1932年4月11日	1933年1月	额定300万元			法租界8号路(今赤峰道)	天津	1932年4月	股本不足,奉令停业
山东民生银行	办事处或代办处	1937年11月	额定600万元 实收320万元	董事长兼总经理 王向荣			济南	1932年7月	查其8个办事处并无天津
国华银行	1934年8月15日	1952年12月15日	额定200万元	董事长 邹敏初 总经理 唐寿民	崔露华	法租界中街74号,1947年迁中正路159号(今解放北路145号)	上海	1928年1月27日	
益发银行	1934年前	1945年	已收20万元			特管区3号路39号	长春	1926年	
益通商业银行	1934年前(办事处)	1945年	额定100万元 实收25万元			特管区3号路39号	长春	1919年1月	
奉天世合公银行	1934年前(办事处)	1936年1月15日	20万元				奉天(沈阳)	1924年	1934年全国银行年鉴总行在长春
中央信托局	1935年10月15日 1945年11月16日复业	1937年后撤 1949年1月	1 000万元	理事长 孔祥熙 局长 张嘉	恽思	1区中正路97号(今解放北路103号)	上海	1935年10月	
宁夏省银行	1935年12月(办事处)		额定200万元 先收四分之一	行长 张承勋		法租界峻庐公寓内	银川	1931年1月	
北平农工银行	1936年1月(办事处)		额定20万元 收足15万元		武向晨	法租界1号路与4号路转角(今大沽路与滨江道)	北京	1935年	发行铜元票
天津市市民银行	1936年4月1日	1949年1月接管	额定100万元 实收50万元	董事长 常筱川(鸿钧)	纪华(仲石)	东北角单街号,1946年迁罗斯福路286号(今和平路)	天津	1936年4月	
满洲中央银行	1937年9月1日	1945年10月	3 000万元	总裁 荣厚		日租界北街85号,后迁兴亚2区3号路86号(今解放北路)	新京(长春)	1931年7月1日	
蒙疆银行	1937年12月(办事处)	1945年10月	额定1 200万元 已收300万元			东马路,后迁日租界旭街15号(今和平路)	张家口	1937年11月23日	
冀东银行	1938年3月10日 1942年7月1日改组	1945年10月	额定500万元 实收250万元	董事长总经理 夏运生		日租界,后迁法租界8号路117号(今赤峰道)	通州,后改唐山,1942年移天津	1936年11月,1942年7月1日改组	
中国联合准备银行	1938年3月10日	1945年8月	额定5 000万元 实收1 250万元	总裁 汪时	唐卜年	北马路,1942年迁法租界中街(今解放北路86号原汇丰银行旧址)	北京	1938年3月10日	发行联银券
新生银行	1942年5月1日	1945年10月	100万元	董事长 陆秀山		法租界26号路18号(今滨江道)	天津	1942年5月	
信诚银行	1942年10月7日	1945年10月	100万元	董事长 王锡桓	周挹清	法租界杨福萌路34号	北京	1942年7月2日	由银号改组成立
华北商工银行	1942年11月16日	1945年10月	额定200万元 实收100万元	董事长 方旭东	张文桂	法租界4号路92号(今滨江道)	北京	1942年11月3日	
华北储蓄银行	1943年3月1日	1945年10月	100万元	常务董事 郑宗圭	饶鸣楷	法租界4号路(今滨江道)	北京	1943年3月1日	
功成银行	(办事处)	1945年10月				11区大安街16号	长春		
唐山农商银行	1943年8月2日	1945年10月	额定200万元 已收100万元	董事长 吴杞芳	赵佩璋	法租界25号路87号(今新华路)	唐山	1943年3月1日	
福顺德银行	1944年	1945年				法租界1号路35号(今大沽路)	烟台		
聚义银行	1944年7月	1945年	100万元			法租界21号路83号(今和平路)	北京	1944年	由聚义银号改组成立
同德银行	1944年9月	1945年				法租界26号路10号(今滨江道)	北京	1944年	由同德银号改组成立
华北工业银行	1944年10月16日	1945年10月	2 000万元 实收半数	董事长 章仲和	史道	新华大楼内(今解放北路,滨江道口)	北京	1944年10月16日	
中国农民银行	1945年12月17日	1949年1月	额定1 000万元 实收500万元	总经理 徐继庄	许锦绶	1区罗斯福路191号(今和平路新华书店)	上海 1946年迁南京	1935年4月	特许发行法币
邮政储金汇业局	1945年12月24日	1949年1月			吕莲渠	1区中正路89号(今解放北路)	上海 1946年迁南京	1930年3月	
亚西实业银行	1946年5月3日	1950年3月29日	15 000万元		姚伯言	1区罗斯福路213号(今和平路)	重庆	1941年1月	
长江实业银行	1946年6月24日	1950年5月2日	200万元			1区罗斯福路133号(今和平路)	重庆 1943年迁昆明	1941年7月15日	
川康平民商业银行	1946年7月3日	1950年3月20日		董事长 刘航琛 总经理 宁芷村	郭秉毅	1区罗斯福路177号(今和平路)	重庆 1947年迁上海	1937年9月	
四川巴川银行	1946年8月15日	1949年1月	2 000万元		白铭五	1区罗斯福路347号(今和平路)	四川铜梁	1940年11月	
大同银行	1946年10月3日	1949年1月	1 000万元		苗作新	1区中正路95号(今解放北路)	重庆 1947年移上海	1943年	
开源银行	1947年1月	1950年3月1日	5 000万元		任熙亭	1区花园路11号	重庆	1941年10月	
亿中企业银公司	1947年3月1日	1949年1月	100万元		程鉴三	1区赤峰道51号	上海	1935年	

近代天津银行一览表
A TABLE OF ALL THE MODERN BANKS IN TIANJIN

银行名称	津行成立日期	津行停业日期	最初资本额	首任负责人	津行经理	津行地址	总行所在地	总行成立日期	备注
中央合作金库河北省分库	1947年3月21日	1949年1月接管	6 000万元	理事长 陈果夫 总经理 袁勉成	陈长兴	1区罗斯福路302号（今和平路）	南京	1946年11月1日	
永利银行	1947年4月	1950年5月4日	1 000万元		陶稚农	1区哈尔滨道112号	重庆1946年迁汉口	1943年	
重庆商业银行	1947年4月29日	1950年7月19日	500万元		缪钟彝	1区中正路113号(今解放北路)	重庆	1930年9月	
中国侨民商业银行	1947年5月	1949年1月	1 000万元		武殿臣	1区花园路8号	昆明	1943年1月	
建业银行	1947年6月14日	1952年12月15日	1 000万元	董事长 汪代玺 总经理 范鸿畴	蔡宝儒	10区营口道安利大楼3号,1950年9月迁解放北路96号(今90号)	重庆 1947年移上海	1944年6月	
敦华商业银行	1948年4月	1949年1月		董事长 张清源 总经理 李松亭	封子平	10区解放南路197号	天津	1948年4月	
联合商业储蓄信托银行	1948年8月1日	1952年12月15日	120亿元	理事长 吴鼎昌 总经理 戴立庵	胡仲文	10区解放北路145号(今147号)	上海	1948年8月1日	四行储蓄部信托部合并改组
汇丰银行（英）	1881年	1941年停业 1945年复业 1954年撤离	港币500万元			英租界宝士徒道(今营口道),1925年迁英租界中街新楼(今解放北路86号)	香港	1864年	发行纸币
德华银行（德）	1890年	1920年停业 1926年复业 1945年8月接管	460万提耳 实收146万			英租界河坝道,1906年迁英租界中街,复业后租用大道(今大沽路)领事道(今大同道)怡和大楼营业	董事会在柏林 总行在上海	1889年	发行纸币
麦加利银行（英）	1895年12月	1941年停业 1945年复业 1954年撤离	300万英镑			英租界中街79号（解放北路153号）	伦敦	1853年	发行纸币
华俄道胜银行（中俄）	1896年	1926年9月	500万两 600万卢布			英租界中街(今解放北路121号)。(1911年该行与法北方银行合并,改称俄亚银行,中文名未改)	彼得堡 1917年改巴黎	1896年	发行纸币
横滨正金银行（日）	1899年	1945年8月接管	1亿日元			英租界中街,1901年迁英租界中街2号(今解放北路80号)	横滨	1880年	发行纸币
华兴银行（中法）	1904年		400万两 中法各半				天津	1904年	
华比银行（比）	1906年	1941年停业 1945年复业 1956年歇业				初在英租界怡和道,1922年迁入英租界中街86号(今解放北路104号)	布鲁塞尔	1902年	发行纸币
东方汇理银行（法）	1907年	1956年歇业	800万法郎			法租界西宾馆。1908年在法租界中街73号建新楼(今解放北路77号),1912年迁入	巴黎	1875年	发行纸币
仪品放款银行（法比）	1907年	1940年	415万法郎			法租界中街111号（今解放北路）	布鲁塞尔	1907年	原名法比银行1910年改现名
天津商工银行（日）	1912年	1920年				日租界寿街(今兴安路)	天津	1912年	1920年与北京实业银行合并组成"天津银行"
正隆银行（日）	1915年	1945年	16万元			日租界旭街25号(今和平路)	营口后改大连	1906年	
中法实业银行（中法）	1916年1月	1921年7月	1 000万法郎			法租界中街西宾馆（今解放北路总工会旧址）	巴黎	1913年7月	发行纸币1923年改为中法工商银行
花旗银行（美）	1916年	1941年停业 1945年复业 1949年歇业				英租界中街通济洋行旧址,1921年迁入其对面新楼即中街66号(今解放北路90号)营业	纽约	1812年	发行纸币
运通银行（美）	1917年	1941年12月	600万美元			法租界7号路39号(今解放北路),后迁入英租界中街137号(今解放北路)victoria Rd	纽约	1841年	
万国储蓄会（法）		1935年	规元4万两先收半数			法租界中街(今解放北路)	上海	1912年8月	
友华银行（美）	1918年	1924年	200万美元			法租界4号路(今滨江道),1924年2月29日归并花旗银行	纽约	1918年	发行纸币
朝鲜银行（日）	1918年	1945年8月	额定4 000万日元 实收2 500万日元			法租界中街93号,1936年迁日租界旭街(今和平路129号[旧191号]新华书店	朝鲜京城	1911年	在华发行纸币
天津银行（日）	1920年	1945年8月	额定250日元 实收四分之一			日租界旭街55号(今和平路)	天津	1920年	
天津实业银行（中日法）	1920年1月		400万元					1920年1月	
华义银行（中意）	1920年5月6日	1940年	华币480万元 英镑64万镑 已收四分之一	华董 许世英		法租界中街91号(今解放北路)	天津 1924年迁上海	1920年5月	1924年中资撤出,改为意外商银行
中华懋业银行（中美）	1920年8月10日	1929年11月	额定1000万美元 实收500万美元	总理 钱能训	张志澄	法租界中街18号(今解放北路)	北京	1920年2月 一说：1919年4月	1921年特许发行纸币

近代天津银行一览表
A TABLE OF ALL THE MODERN BANKS IN TIANJIN

银行名称	津行成立日期	津行停业日期	最初资本额	首任负责人	津行经理	津行地址	总行所在地	总行成立日期	备注
华法银行(中法)	1920年10月	1931年	2 000万法郎	华总裁 王鸿陆		法租界8号路111号(今赤峰道)	天津	1920年10月	
远东银行(意)	1921年	1929年8月16日	额定5 000万里耳 先收1 250万里耳			意租界大马路36号(今建国道)	上海	1921年	
振业银行(中法)	1921年	1924年9月破产	额定500万元 先收200万元	总董 张寿龄			北京	1917年2月	自行发行纸币
震义银行(中意)	1921年5月	1924年	1 000万元				北京	1921年5月	发行纸币
大东银行(中日)	1922年	1927年4月	额定500万元 实收250万元			日租界旭街2号(今和平路)，后迁法租界中街(今解放北路)	北京 后改上海	1921年5月	
美丰银行(美)	1923年	1935年5月24日	额定731万美元 实收389万美元			法租界中街(7号路)61号(今解放北路)	上海	1917年	发行纸币
中华汇业银行(中日)	1924年3月	1928年12月	额定1 000万日元 实收500万日元	总理 陆宗舆		法租界海大道(今大沽路)34号，1927年迁英租界中街9号(今解放北路117号)	北京 1927年迁天津	1918年2月	发行纸币
中法工商银行(中法)	1925年11月	1948年末	2 000万法郎			法租界中街114号(今解放北路76号)	巴黎	1923年	由中法实业银行改组更名
远东银行(俄)	1925年	1929年9月	500万卢布			法租界大沽路	哈尔滨	1923年	
汇源银行(法)			额定100万元 实收50万元			英租界29号路387号(今南京路)	上海	1921年	
汇源信托银行(法)						法租界狄总领事路66A号(今哈尔滨道)	上海		
天津商业放款银行(美)	1928年(俄) 1932年改组(美)	1936年1月	1万元，1932年改组后资本金2.5万美元			英租界大沽路245号	天津	1928年由旅津俄人创办，1932年改组向美国政府注册	
大通银行(美)	1929年 1946年复业	1941年停业 1949年1月	美金500万元			英租界中街1号(今解放北路113号)，后迁英租界中街80号(即10区中正路108号，今解放北路100号)	纽约	1920年 1931年合并改组	
合通银行(美)	1933年11月 1946年复业	1935年8月停业 1949年1月	美金10万元		华经理 杜贯三	法租界巴斯德路即8号路113号，1947年迁中正路137号(今解放北路)，1948年2月迁中正路181号(今解放北路)	天津	1933年 (一说1935年4月)	由合通洋行改组成立
敦华银行(美)	1935年12月28日 1946年2月复业	1941年停业 1949年1月清理	2万美元			英租界中街177号，复业后在中正路197号(今解放北路)	天津	1935年	

下列银行在天津只有代理关系并无分支机构

哈尔滨辅商银行：总行设在哈尔滨，天津未设分行。参见《银行周报》1919年前后该行广告。

广 东 银 行：总行设在香港，成立于1912年，天津未设分行。参见《银行周报》1919年前后该行广告。

东 亚 银 行：总行设在香港，成立于1919年，资本港币1 000万元，实收500万元，天津只设代理关系。参见《银行周报》1926~1931年该行广告。

另有1920年成立的总行设在菲律宾马尼拉的"中兴银行"，资本1 000万批沙，实收560万，只设分行于上海、厦门，天津未设机构，只是与天津的"中兴银行"同名。1934年在上海成立的"大康银行"无分支机构，《天津商会档案汇编》所列的"大康银行"无内容，未找到任何线索，天津有大康银号。故未列。

注：以上"近代天津银行一览表"系根据《全国银行年鉴》、《银行周报》、《银行月刊》、《天津商会档案汇编》、《中国储蓄银行史》、《大陆月刊》、《中行月刊》、《中联月刊》、各银行档案全宗说明等有关银行资料和《天津通志·大事记》等汇集，经过比较、考证、选择、整理而成。对资力较小、开设时间不长的银行因缺乏资料，只在备注栏内注明出处备查。

1937年天津银号名录
NAMES OF TIANJIN MONEY SHOPS IN 1937

字 号	帮派	业 务	所在地	字 号	帮派	业 务	所在地
信源溢银号	本地帮	折交，浮事	日租界荣街	庆成银号	本地帮	折交	宫南
余大亨银号	本地帮	折交，浮事，汇兑	法租界5号路	广瑞银号	本地帮	折交	法租界杨福荫路
中宝银号	本地帮	折交，浮事	日租界旭街	广源银号	本地帮	门市，票行	法租界梨栈大街
宝生银号	本地帮	折交，浮事	日租界旭街	源合春银号	本地帮	折交，门市	特1区芝罘路
天瑞银号	本地帮	折交，浮事，汇兑	法租界5号路	源达银号	本地帮	折交，浮事	宫北
宏康银号	本地帮	折交，浮事。汇兑	宫北福神街	万溢银号	本地帮	折交，浮事	法租界6号路
同裕厚银号	本地帮	折交，汇兑	锅店街	顺兴银号	本地帮	浮事，门市，票行	法租界梨栈大街
颐和银号	本地帮	折交，浮事，仓库	法租界26号路	孚庆银号	本地帮	浮事，门市，票行	法租界梨栈大街
裕源银号	本地帮	折交，汇兑	针市街	源泰银号	本地帮	折交，浮事	宫北
中和银号	本地帮	折交，汇兑	法租界6号路	万昌银号	本地帮	折交，票行	法租界
和丰银号	本地帮	折交，浮事，汇兑	法租界6号路	裕大银号	本地帮	浮事，门市	日租界旭街
万华银号	本地帮	折交	法租界1号路	祥发银号	本地帮	浮事，汇兑	日租界福岛街
元泰银号	本地帮	折交，浮事，汇兑	宫北	永昌银号	本地帮	折交	
泰丰恒银号	本地帮	折交，浮事	宫北	老恒利银号	本地帮	折交	日租界旭街
天兴恒银号	本地帮	折交	河北大街	桐丰银号	本地帮	浮事，折交	法租界杨福荫街
义胜银号	本地帮	折交，汇兑	法租界4号路	振兴长银号	本地帮	折交	法租界杨福荫街
义恒银号	本地帮	折交	东门外	大昌银号	本地帮	折交	法租界5号路
晋生银号	本地帮	折交，汇兑	法租界30号路	大康新记银号	本地帮	浮事	日租界旭街
敦昌银号	本地帮	折交，浮事，汇兑	宫北	陆泰银号	本地帮	浮事	法租界24号路
益兴珍银号	本地帮	折交，浮事	法租界杨福荫路	津源银号	本地帮	浮事，汇兑	日租界
泰和银号	本地帮	折交	针市街	祥丰银号	本地帮	浮事，门市	日租界旭街
瑞源永银号	本地帮	折交，浮事	法租界	永生银号	本地帮	折交	法租界4号路
谦义银号	本地帮	折交，汇兑	法租界23号路	锦生银号	本地帮	浮事，门市	日租界旭街
福康仁银号	本地帮	折交	法租界	中兴银号	本地帮	折交，汇兑	针市街
永同生银号	本地帮	折交	法租界30号路	中裕银号	本地帮	折交	针市街
义成裕银号	本地帮	折交，浮事，汇兑	北门外	诚明银号	本地帮	折交，汇兑	针市街
振义银号	本地帮	折交	法租界5号路	利生银号	本地帮	浮事，票行	日租界旭街
谦丰银号	本地帮	折交	法租界	利丰银号	本地帮	门市	日租界旭街
庆益银号	本地帮	折交	法租界4号路	庆昌银号	本地帮	浮事	日租界曙街
祥生银号	本地帮	折交	法租界4号路	德源厚银号	本地帮	门市，浮事	日租界旭街
德仁银号	本地帮	折交	法租界30号路	永泰银号	本地帮	折交	法租界23号路
辉远银号	本地帮	折交	法租界5号路	玉盛银号	本地帮	门市	法租界4号路
庆瑞银号	本地帮	折交	竹竿巷	福康银号	本地帮	浮事	法租界32号路
宏源银号	本地帮	折交	竹竿巷	聚元银号	本地帮	票行，门市	法租界
华丰银号	本地帮	折交	法租界20号路	民丰银号	本地帮	门市	英租界小白楼
余庆银号	本地帮	折交，浮事	法租界27号路	信丰仓库银号	本地帮	折交（抵押放款）	英租界海大道

1937年天津银号名录
NAMES OF TIANJIN MONEY SHOPS IN 1937

字 号	帮 派	业 务	所在地
聚丰银号	本地帮	门市	英租界
福泰银号	本地帮	门市	特3区大经路
积昌银号	本地帮	折交	针市街
庆亿银号	本地帮	浮事	日租界旭街
永信银号	本地帮	折交	法租界4号路
义诚厚银号	本地帮	浮事	法租界14号路
兴源银号	本地帮	票行，门市	法租界梨栈大街
庆生银号	本地帮	折交，汇兑	日租界旭街
久大银号	本地帮	浮事	日租界宫岛街
永增和银号	京 帮	折交，浮事	法租界27号路
谦生银号	京 帮	折交	法租界40号路
聚义银号	京 帮	折交，汇兑	针市街
敦泰永昌记银号	京 帮	折交，浮事	法租界1号路
致昌银号	京 帮	折交，汇兑	法租界20号路
聚泰祥银号	京 帮	折交	法租界23号路
同德银号	京 帮	折交，仓库	法租界26号路
广业银号	京 帮	折交，汇兑	法租界杨福荫路
庆聚银号	京 帮	折交	法租界文兴里
福记银号	京 帮	折交	法租界
聚华昌银号	京 帮	折交	法租界
源通厚银号	京 帮	门市，折交	法租界1号路
隆远银号	京 帮	折交	法租界30号路
同兴银号	京 帮	折交	法租界
生生银号	京 帮	折交	北门西
太和记银号	京 帮	折交	法租界24号路
祥瑞兴银号	京 帮	折交	法租界23号路
致远银号	京 帮	汇兑	法租界
春兴银号	京 帮	折交	法租界
万义长银号	京 帮	折交，汇兑	北大关
宏义银号	京 帮	折交	针市街
慎兴银号	京 帮	折交	北门西
瑞记银号	京 帮	汇兑	日租界宫岛街
聚源银号	京 帮	折交	法租界杨福荫路
正昌银号	京 帮	折交	法租界杨福荫路
春和银号	京 帮	折交	法租界杨福荫路
宏庆余银号	京 帮	折交，汇兑	英租界2号路
恒兴银号	京 帮	汇兑，浮事	法租界25号路
锦记兴银号	京 帮	折交	日租界
义兴银号	京 帮	汇兑	北马路
全记银号	京 帮	折交	法租界23号路
金茂昌银号	京 帮	汇兑，折交	法租界24号路
聚盛源银号	京 帮	折交	英租界2号路
鸿记银号	山西帮	折交，汇兑	北门内
万德银号	山西帮	折交	法租界30号路
立昌永银号	山西帮	折交	法租界29号路
蚨亭银号	山西帮	折交，浮事	法租界5号路
元吉银号	山西帮	汇兑	针市街
亭记银号	山西帮	汇兑	法租界32号路
宏业银号	山西帮	汇兑	法租界梨栈
益昌银号	山西帮	汇兑	法租界4号路
汇源永银号	山西帮	汇兑，折交	法租界25号路
兴盛永银号	山西帮	折交，汇兑	法租界梨栈
永瑞银号	山西帮	汇兑	英租界2号路
冀鲁银号	山东帮	折交	法租界27号路
启明银号	山东帮	折交	法租界25号路
德祥银号	山东帮	汇兑	法租界梨栈
同增益银号	山东帮	汇兑	法租界31号路
寿康银号	山东帮	折交，汇兑	英租界11号路
文兴银号	东 帮	折交	法租界海大道
万兴长银号	东 帮	浮事，汇兑	法租界
恒泰银号	东 帮	浮事，汇兑	法租界杨福荫路
福顺银号	东 帮	汇兑	北马路
和丰裕银号	河南帮	汇兑	法租界6号路
恒源益银号	河南帮	汇兑	法租界6号路
华兴银号	河南帮	汇兑	法租界海大道
同裕银号	南宫帮	折交	法租界27号路
恒利银号	南宫帮	折交	英租界广东路
聚德银号	河间帮	折交	日租界秋山街
裹通银号	河间帮	折交	法租界32号路
裕通银号	热河帮	汇兑	特3区2经路

□ 罗澍伟 / Luo Shuwei

老银行 — 天津近代经济的核心
Old Banks – Central to Tianjin's Economic Growth

昔日天津海河
Tianjin's River Hai in the old days

如果能把历史的时针倒旋至19世纪末叶，你就不难发现，在渤海之滨、海河之畔，已经有一座银行林立的城市，这就是开埠以后的天津。

由于天津地处南北运河与海河交汇处，自明清以来，一直是一个粮、棉和土特商货集散的码头，五方杂处，物阜人丰，商业贸易的发展，自然带动了银钱业的发展，钱庄、银号应运而生。

18世纪以前的商业贸易，无论大宗小宗，都使用现金交易。19世纪初，由于社会动乱，沿途不靖，客商转运现银，时有被劫夺的现象，因此汇兑业迅速发展。据说在19世纪初，天津有一家山西人开设的颜料铺分号，去四川采买铜绿时，款项先由四川分号垫付，货到后再由天津分号结账，十分方便。从此不少商家纷起效尤，结果竟成为中国汇兑业之始。

1860年，天津被迫开放为通商口岸。开埠以后，因为有着华北、东北和西北广大地区做腹地，天津的内外贸易异常兴旺，极大刺激了天津金融业的发展。到19世纪末，天津的钱庄、票号已达200余家。

无论钱庄也好，票号也好，一般只能服务于本地区或埠际商业贸易；特别是近代以前，中国的进出口贸易都是以现银交易，没有所谓国际结算。近代以后则不同了，进出口贸易的结算与支付，因世界各国的币制不同、单位不同，必须通过一体化的现代银行来运行，这种接轨对于一个已经成为国际化城市的天津来说是必不可少的。

到了19世纪80年代，天津的进出口贸易已具相当规模；但对于进出口商来说，贸易资金的提供只能依靠一些有实力的洋行附设的银行业务部来进行。当时，香港是英国对中国各通商口岸进行进出口贸易的基地，然而香港与中国各口岸之间没有直接的金融联系，因此筹建一家总行在香港、分行设在各口岸的现代银行势在必行。

昔日解放桥繁忙景象
Busy scene of Jiefang Bridge in the old days

1865年，汇丰银行香港总行和上海分行同时成立，它的英文原名就是Hongkong & Shanghai Bank。1880年汇丰银行开始筹设天津分行，翌年正式开业，主要业务是向资金短缺的中国提供建设贷款，比如天津到山海关的津榆铁路就是依靠汇丰贷款修建的。

在当时天津的对外贸易中，以中英间的贸易额最大，汇丰天津分行的成立，使天津港的贸易结算无须再经过上海，从而降低了贸易成本，提高了竞争力。19世纪的英国，在全球贸易中居于领军地位，汇丰银行建立不久，就把分行扩展到欧美各国。汇丰分行是天津最早建立的外资银行，较早掌握了天津的外汇市场。

1890年由德国几家大银行投资设立的德华银行天津分行，成为天津的第二家外资银行。德华银行总行设在上海，但董事会却在柏林。德华银行天津分行最初为代理店性质，由于与德国政府关系密切，在天津德租界划定之初，德国政府的一切开辟和建设工作均委托德华银行代理。

北洋天津银号发行的纸币
Banknotes issued by Beiyang Tianjin Bank

1895年英国另一家皇家特许银行——麦加利银行天津分行开业。麦加利银行也叫渣打银行，总行设于英国伦敦。主要业务是定活期存款、汇兑信用证；设有保管部和信托部，代客保管一切物件；经营募集债券，发行和买卖证券等业务。为适应中国的金融市场，该行发行了银两和银元两种兑换券。

1896年另一家外资银行——华俄道胜银行天津分行开业。这家银行的总行设在彼得堡，名为华俄，实际上是由法国和俄国共同投资。除了京、津、沪、汉四大城市，道胜银行的20多家分行遍布中国东北。鉴于20世纪初天津经济地位的不断上升，1910年道胜天津分行还成立了外币交易所。

1898年总行设于法国巴黎的东方汇理银行天津分行开业。该行在越南西贡设有总监理处，统辖远东各地分支机构。天津分行主要经营进、出口押汇及外汇买卖。

1899年总行设在日本横滨的正金银行天津分行开业。该行以"清算中日两国间贸易汇兑为主要业务"为幌子，实际上是一家服务于日本扩张和侵略政策的对外贸易银行。

1916年花旗银行天津分行开业。该行总行于1812年成立于美国纽约，天津分行独家负责美国对华贸易输出结算，是在津美资银行势力最大的一家，除办理存放款业务外，还发行钞票。

北洋天津银号
Pei-Yang Tientsin Bank

20世纪初到30年代中，在天津设立的外资银行又有十几家，其中包括美国的运通银行、美丰银行、大通银行、合通银行与敦华银行，比利时的义品放款银行与华比银行，法国储蓄银行（万国储蓄会）与中法工商银行，日本的正隆银行与朝鲜银行等。

在外资银行迅速发展的同时，从20世纪初开始，随着工商业和口岸贸易的发展，中国人自己经营的银行和非银行金融机构也在天津大规模地发展起来。当时全国有20多家著名的银行先后在天津设立总行或分行，天津的多家银行也在全国各地建立了分支机构。

属于国家银行的有1904年在天津北马路设立的户部银行天津分行。1908，户部改为度支部，户部银行相应改为大清银行。辛亥革命后，大清银行被清理，1912年改组为中国银行，是中华民国的中央银行，同

昔日商业街
The former business street

年在天津设分行，成为该行北方地区的管辖行。

交通银行于1908年设天津分行。第一次国内革命战争后，中央银行于1931再次设天津分行。中国农民银行于1945年设天津分行。

属于地方性银行的有1902年成立的直隶省官银号，同时在北京、上海、汉口、保定、张家口、唐山设分号，1910年改为直隶省银行，1929年改组为河北省银行。天津市民银行，成立于1936年，日伪统治期间一度成为伪联合准备银行下属的银行。直东地方实业银行，1924年成立，总行设在天津。此外还有山西裕华银行，1915年成立，总行设山西太谷，1927年迁天津。

省级银行中，北京中华储蓄银行1924年前设立天津分行。北京商业银行，1927年设天津分行。

山东工商银行，系山东省银行，1925年设天津分行。山东民生银行于1927年设天津办事处。河南省银行及河南省农工银行，均于1929年设天津分行。

东三省银行，1920年设天津分行。奉天商业银行于1924年设天津分行。西北银行于1926年设天津分行。山西省银行于1927年设天津分行。陕西省银行、甘肃省银行和宁夏省银行，则分别设有天津办事处。

属于官商合办的有中国通商银行，1898设天津分行。中国实业银行，1919成立于天津。新华信托储蓄银行于1917年设天津分行。中国国货银行于1931年设天津分行。

属于商业银行的，有1915年成立的盐业银行，1928年总行迁往天津。1917年成立的金城银行，总行设天津，1936年迁上海。成立于1919年的大陆银行，总行设天津。1921年成立的中南银行，总行设上海，1922年设天津分行。1922年，盐业、金城、大陆、中南联合成立四行准备库及四行储蓄会。这四家银行资金雄厚，其中盐业与中南的总行虽分设在北京和上海，但这两家银行的股东大都居住在天津，并且经营的重点也在天津，所以这四家银行并称为"北四行"，其金融实力甚至超过了总行设在上海的浙江实业银行、兴业银行和商业储蓄银行等"南三行"。

其他的商业银行或非银行金融机构，有1916年成立的中孚银行，总行设天津，1930迁上海，天津设分行。上海商业储蓄银行，1920年设天津分行。中国农工银行，1927年设天津分行，1929年总行迁往天津，1931迁上海。中国垦业银行，1926年成立，总行设天津，1929年迁上海。

值得注意的是，天津传统的钱庄与票号并未因银行业的快速发展而势微，它们紧跟天津经济发展之所需，疏通了各种融资渠道，因此同样得到了空前的发展。到20世纪30年代，天津的银号总数竟占到北平（北京）、天津、济南、青岛四大城市的一半以上，而资金总额更占到三分之二。在以天津为中心的金融网络之中，钱庄、票号发挥了不容忽视的巨大作用。

旧时天津街景
Tianjin Old streets

所以说，当时的天津，已经成为中国北方最大的工商业和港口贸易城市，也是全国金融业最发达、金融交易最繁忙的城市之一。

当前，天津已被国家定位为北方经济中心和国际港口城市，滨海新区开发开放提升到国家战略高度。这极利于把天津建设成为中国金融业改革开放和试验创新基地，中国金融业与世界金融业最便捷的接轨点，续写天津金融新的辉煌，同时天津也面临着最好的历史机遇。

If we could turn back time to the end of 19th century, we would see Tianjin as both a major Chinese trading port and financial center.

Tianjin as a trading city was ideally located at the convergence of the Grand Canal and the Haihe River, the main trade route for goods to Beijing. It served as a distribution port for important commodities such as grain and cotton throughout the Ming and Qing Dynasties. With a large population from all regions of China its products were extremely diverse. This influx of trade triggered the development of its banking business and led to the emergence of banking and financial institutions.

Prior to 18th century business transactions were all conducted in cash. In the early 19th century, social unrest, made it unsafe for merchants to transfer cash. As a result, remittance business developed rapidly in Tianjin. It is said that a businessman from Shanxi established a dye shop branch in Tianjin. When he went to Sichuan to buy verdigris dye, his branch in Sichuan paid for the purchase first. Later when the goods arrived in Tianjin, the Tianjin branch would settle the account. As it was so convenient, many other shops followed suit. That was the beginning of China's remittance business.

In 1860, Tianjin was forced to open as a foreign trading port. Since it had North China, Northeast China and West China as its hinterlands, domestic and foreign trade in Tianjin flourished. This rapidly increased the demand for financial services. By the end of 19th century, there were as many as 200 banks and other financial institutions in Tianjin.

卞白眉（1884-1968）江苏仪征人。1906年考入美国白朗大学，取得文学士学位。于1912年回国，参与筹办中国银行，1913年中国银行正式成立，成为当时政府的中央银行。卞白眉入行之初，任发行局佐理，后升任总稽核。1916年辞职移居天津，协助孙多森筹办中孚银行。在天津金融界，卞白眉先后工作了20个春秋。

Bian Baimei, also known as Shou Sun, was born in Yizheng, JiangSu in year 1884. In 1906, Bian Baimei went to the United States of America to read Politics and Economics at Brown University, where he graduated with a Bachelor in Philosophy. He returned to China after the Xinghai Revolution and started his career in the Chinese financial market. He was the general manager and assistant manager of the Tianjin Branch of the Central Bank. Prior to that, he worked as an assistant and auditor in chief for the Monetary Policy Division of the Central Bank. Bian Baimei made significant contributions to financial market during his twenty years working for the Tianjin Branch of the Central Bank.

昔日的中国银行天津分行旧址
The former Bank of China, Tianjin Branch

第一代汇丰银行天津分行旧址
The former HSBC Tianjin Branch

旧时的金城银行
天津分行外景
Old Kinchen Bank Tianjin Branch

These banks were only able to provide services for trade between Chinese cities. All imports and exports had to be settled in silver dollars. With the advent of modern banking, things changed. Settlement of, and payment for import and export trade was handled by merchant banks, using currency exchanges. It was essential for Tianjin, as a trading city, to follow.

By the 1880s, import and export trade of Tianjin had developed to a considerable scale. However, the only providers of trade funds for import and export transactions were powerful foreign bank. At that time, Hong Kong was a base for Britain to trade with Chinese ports; but there was no direct financial link between Hong Kong and the mainland ports. It became imperative to set up a modern bank with its head office in Hong Kong and branches dispersed in all the trading ports.

In 1865, the Hong Kong & Shanghai Banking Corporation was established with its head office in Hong Kong and a branch office in Shanghai. In 1880, HSBC as it is known began preparation for the establishment of a branch in Tianjin. The following year it commenced formal operation. The principal business of Tianjin branch was to provide construction loans to China. For example the Jin-Yu railway from Tianjin to Shanhaiguan was built with HSBC loans.

Of all the foreign trade in Tianjin, the trade volume between China and Britain was the greatest. The establishment of HSBC Tianjin branch lowered trading costs and improved the competitive power of Tianjin. Transactions concluded in the port of Tianjin no longer needed to be settled via Shanghai. Britain in the 19th century occupied a leading position in world trade. Therefore, not long

金城银行行长周作民，银行家。1917年5月创办金城银行，任总经理。金城银行，曾为全国私营银行之首。1951年9月，任公私合营的"北五行"董事长，1952年12月，任统一的公私合营银行副董事长。

The Chairman of Kinchen Bank, Zhou Zuomin, was born in Weian Jiangsu in 1884. He founded Kinchen Bank in May 1917 and was appointed as the General Manager of the bank. Kinchen Bank was the foremost private bank in China. In September 1951, he became the chairman of the "Alliance of the Five Northern Banks", which was jointly owned by the state and private individuals.

第一代横滨正金银行
天津分行旧址
The business hall of the Yokohama Specie Bank, Tianjin Branch

after its founding, HSBC started to establish branches in countries and regions in Europe and America. HSBC Tianjin branch, being the first foreign bank, acquired a monopoly over foreign exchange markets in Tianjin at an early stage.

In 1890, several major German banks set up the Deutsche Asiatische Bank Tianjin Branch; This was only the second foreign bank in Tianjin. The bank had its head office in Shanghai and its board of directors in Berlin. Deutsche Asiatische Bank Tianjin branch was initially a correspondence bank in nature. Since it enjoyed close relationship with the German government, the government entrusted it with the development and construction of German concessions in Tianjin.

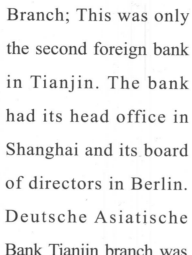

昔日的东莱银行天津分行
The former Tung Lai Bank, Tianjin Branch

德华银行旧址
The former building of the Deutsch Asiatische Bank

In 1895, another British bank--Chartered Bank of India, Australia & China Tianjin branch went into operation. Chartered Bank of India, Australia & China, also known as Standard Chartered, had its head office in London. It offered deposit-taking and credit remittance services. It had a safe deposit department and a trust department which took care of all types of articles for its clients. It also engaged in raising bonds, in issuing, buying and selling securities.

To meet the demands of China's financial market, the bank issued two kinds of exchange certificate: silver certificate and silver dollar certificate.

In 1896, another foreign bank the Russo-Chinese Bank Tianjin branch was founded and went into business. This bank, with its head office in St. Petersburg was in fact a joint investment by France and Russia. In addition to branches in the four big cities of Beijing, Tianjin, Shanghai and Hankou, it also had more than 20 branches all over Northeast China. Seeing that Tianjin was

中法储蓄会旧址
The former building of China-France Savings Bank

improving in its economic status in the early 20th century, the Bank founded a foreign exchange in 1910.

In 1898, L'indo Bank of China Tianjin branch started business. It had its head office in Paris, France and a general supervision office in Saigon, Vietnam which was in charge of all the branches in the Far East. Tianjin branch was mainly responsible for import and export loans and foreign exchange dealings.

In 1899, the Yokohama Specie Bank Ltd. Tianjin branch went into business. The Yokohama Bank was a Japanese bank with its head office in Yokohama. This bank was founded primarily in order to settle trade transactions between China and Japan.

In 1916, The International Banking Corporation put its Tianjin branch into

交通银行天津分行旧址
The former building of the Bank of Communications, Tianjin Branch

operation. Its head office in New York, U.S. was set-up in 1812. The Tianjin branch had sole responsibility for the settlement of America's export transactions with China. It was the most influential of all American banks in Tianjin. Besides deposit-taking and lending, it also issued bank notes.

From early 20th century to 1930s, many more foreign banks established branches in Tianjin. Among them, were American Express Co, Inc, American Oriental Banking Corporation, Equitable Eastern Banking Corporation, Credit Foncier d' Extreme Orient and Banque Belge Pour L'Etranger from Belgium, French Savings Bank, Banque Francochinoise Pour le Commerce et l' Industrie , Japanese Zhenglong Bank and Korean Bank.

At the time when foreign banks experienced rapid expansion in China, banks and other financial institutions run by Chinese people themselves began to develop on a large scale. In Tianjin industry, commerce and trade all expanded during the early 20th century. Nationwide more than 20 well-known banks set-up their head offices or branches in Tianjin. Many Tianjin banks

横滨正金银行天津分行办公大厅
The business hall of the Yokohama Specie Bank Tianjin Branch

likewise established branches in other parts of China.

One of three state banks that had branches in Tianjin was the Household Ministry Bank. The Tianjin branch opened in 1904, later becoming the Ta Ching Bank when the Household Ministry changed to the Central Ministry of Finance in 1908. The Revolution of 1911 led to the winding-up of Ta Ching Bank. It was reorganized into the Bank of China in 1912 as the Central Bank of the Republic of China. In the same year, its Tianjin branch was set up to direct all the other branches in northern China.

The Bank of Communications founded its branch in Tianjin in 1908. After the First Revolutionary Civil War (Northern Expedition), Central Bank for the second time set up a branch in Tianjin. The Farmers Bank of China set up its Tianjin branch in 1945.

当年使用过的老账本
The old Bookkeeping

二十世纪初的大通银行营业大厅
The business hall of the Datong bank in the early 20th century

Some native banks also had branches in Tianjin. In 1902 Zhili Province set up its own financial institutions. Branches were established at the same time in Beijing, Shanghai, Hankou, Baoding, Zhangjiakou and Tangshan. Its name became Bank of Zhili Province in 1910, and was later reorganized as Bank of Hebei Province in 1929. Tianjin Citizens' Bank was founded in 1936 and was for a time affiliated with United Preparatory Bank of the Japanese puppet government. Zhidong Industrial Bank was founded in 1924 with its head office in Tianjin. Shanxi Yuhua Bank was founded in 1915 with its head office in Taigu, Shanxi. It moved the head office to Tianjin in 1927.

Of the provincial banks, Beijing Zhonghua Savings Bank set up its Tianjin branch before 1924; Beijing Commercial Bank set up its Tianjin branch in 1927.

Many provincial, industrial and commercial bank, mainly from northern China established branches in Tianjin during the next few years. Banks jointly run by government and merchants included China Imperial and Commercial Bank, The National Industrial Bank of China, Xin Hua

Trust and Savings Bank Ltd., and China National Goods Bank also established branches.

The four leading commercial banks, Salt Industrial Bank, Kinchen Bank, Mainland Bank and The China & South Sea Bank Ltd., together established the Northern Four Saving Banking Society in 1922. United they possessed powerful financial strength. Though Yien Yieh Industrial Bank and The China & South Sea Bank Ltd. had their head offices in Beijing and Shanghai respectively, most of their shareholders lived in Tianjin and their business concentrated in Tianjin. That was why the four banks became known as Northern Four Banks. They were even stronger in financial terms than the three main banks from southern China namely Zhejiang Industrial Bank, The National Commercial Bank and Commercial Savings Bank that had their head offices in Shanghai.

What is interesting is that the original Chinese style money houses in Tianjin did not suffer due to the rapid development of its banking business. They opened up diverse financing channels to meet the demands of the economic development in Tianjin. They were rewarded with unprecedented expansion. By the 1930s Tianjin dominated all local financial transactions for the surrounding provinces. It can be said that Tianjin had already become the mightiest industrial, commercial and trade city in northern China.

Today China has designated Tianjin as the northern economic center and international port. The opening up and development of its Binhai New Area has been listed as a national strategic tasks.

Tianjin now has perhaps the best opportunity to renew its financial splendor by building itself into a financial industry base. Through economic reform and allowing greater freedom to the finance industry Tianjin is once again incorporate itself into the global financial market.

东方汇理银行天津分行旧址
The former building of Indo-China Agency Bank Tianjin Branch

沈理源1890年7月11日生于浙江，他于1909年考入意大利拿波里奥工业大学，攻读土木和水利工程，是一位建筑设计大师。

主要设计作品有杭州浙江兴业银行，清华大学体育馆扩建工程，清华大学电气馆、机械馆、航空试验馆和教工住宅，北京大学沙滩图书馆，天津浙江兴业银行，天津盐业银行，天津新华信托储蓄银行。此外，他还设计了不少商业建筑和私人宅邸等。

沈理源先生于1951年11月21日病逝于北京。

Famous Architect Mr. Shen Liyuan was born on 11th July 1890 in Zhejiang. He was admitted to the Università degli Studi di Napoli Federico II, in 1909 to read civil engineering and hydraulics.
His masterworks include the Hangzhou Branch of the National Commercial Bank, the extension of the Tsinghua University Stadium; the Departments of Electronics, Mechanics and aeronautics of Tsinghua University; the Peking University Library; Tianjin Branch of the National Commercial Bank; Tianjin Branch of the Salt Industrial Bank and Tianjin Branch of Xin Hua Trust & Saving Bank. He also designed a number of commercial buildings and private estates.
Mr. Shen Liyuan died on 21st November 1951 in Beijing.

跋
POSTSCRIPT

抚摸"流金"岁月

□ 田贵明

打开这本画册，天津历史上的"流金"岁月扑面而来：

盐业银行厚重的大理石楼梯上，昔日主人的脚步声仿佛正橐橐已远；汇丰银行金碧辉煌的营业大厅里，各国客商的交谈声仿佛余音绕梁；浙江兴业银行巨大的石柱上，当年的抚摸仿佛余温尚存；中法工商银行走廊上的迷离的华灯，仿佛正等待主人远行的归来；四行储蓄会门前冬枝上的瑞雪，仿佛正在温暖的阳光下滴答消融；金城银行油漆斑驳的金库大门，仿佛倦怠了对开启的等待……柱顶上的石雕，扶梯上的木纹，彩窗上的玻璃，穹顶上的彩绘，过道里的铜门……

这里说的"流金"岁月，不是泛指那些金子般流走的逝水光阴，而是特指19世纪末叶20世纪初天津作为北方金融中心，银行林立、金融流通空前活跃的那个年代。

那是天津历史上名副其实的"流金"岁月。

天津1860年首度开埠。内达三江，外通四海，优越的地理位置和便捷的水陆交通条件吸引各国客商纷至沓来。西风东渐，天津的经济生活日趋活跃，贸易汇兑也从传统操作方式渐渐与国际接轨。这其中，银行业的异军突起蓬勃发展成为其中一道亮丽的风景。尤其是外资银行，挟西方资本主义工业革命余绪，开津门现代金融经济之先河。随着内外贸易的不断扩大，一座座银行建筑应运而生。这些来自世界各地的银行机构靠自身内部的积极运作，成为城市现代经济生活重要的组成部分，增强着城市的内部功能，使当时的天津开始逐步走向全国，走向世界。金融流通，岁月流金，天津在中国乃至世界的经济舞台上，开始扮演愈来愈重要的角色。

源于归属国度及主人的不同，这些来自世界各国的银行，在外部观感、内部结构尤其细节设计上，也呈现出个性鲜明、风格独具的建筑风貌，它们错落有致，精美绝伦，极大丰富了这座城市的建筑艺术展台，成为这个开埠城市自身文化突出的形象特征。

可惜的是，由于自然的风雨、战火的涂炭、"文革"的浩劫等原因，这些银行建筑，与这个城市当年引以为傲的风情区建筑一样，在失去其原有的内部功能后，其外部容颜、内部结构有些受到了不同程度的损毁，其历史地位与文化价值亦遭蒙尘。而作为中国当时三大直辖市之一的天津，其城市文化特色也随之失色。

文化学者认为，城市文化特色是一个城市文化积淀的外在体现，是一个城市内在本质的外部表象，由历史沿革、经济发展、地理特征等多重因素共同作用、长期演化而来。一个城市的文化特色越鲜明，就越能赢得世人的赞誉。

在改革开放国际交流日益频繁的今天，在天津与世界拥抱更为紧密的新世纪，重新审视、挖掘、整理、认识、提升这些银行建筑作为珍贵历史遗存的历史价值、文化价值、经济价值、外交价值甚至品牌价值，发扬光大之，使之四海闻名，彰显天津的丰厚历史底蕴与美好城市风貌，为繁荣天津市的经济建设所用，应该是很有必要。一些民间摄影家艺术家的长期关注与积极努力，让我们对这些价值的重新审视、挖掘、整理、认识与提升，变得更为方便。

令人感奋的是，在我们对这些历史遗存做着整理与挖掘的同时，新的历史机遇再次惠顾天津：渤海银行建立；滨海新区被批准为国家综合配套改革试验区；北方经济中心的明确定位；一项项有关开发开放、有关金融改革金融发展的新政不断出台，在天津先行先试；一家家外资银行重新扎根落户，散叶舒枝……

一个金融经济更为繁荣、金融流通更为活跃的新"流金"岁月再次来到天津。金融，作为现代经济的重要支撑，再次成为天津崭新历史舞台的主角。天津，正在重新成为世人瞩目的北方金融之都。

抚摸"流金"岁月，感觉昔日重来。瞩望明日天津，一片金光灿烂。天津，有着"流金"岁月的丰厚历史底蕴，也有着"流金"岁月更为金色的明天。

让我们共同为她祝福吧。

A Touch upon the "Golden Age"

☐ Tian Guiming

The moment you open the album, you will be enchanted by the splendid of history of Tianjin's golden age.

The footsteps of the former proprietors of Salt Industrial Bank seem to be fading away from its heavy wooden staircase; the chatter of clients from all over the world seem to be echoing around the magnificent business hall of HSBC; the flickering lamps over the hallway of Banque Francochinoise Pour le Commerce et l' Industrie seem to be expecting the return of their master from afar; embraced in the warm sunshine, the snow on the twigs in front of the Northern Four Savings Bank seems to be melting; the treasury door of the Kinchen Bank, mottled with paint, seems to be weary of waiting to be reopened…as are the stone sculptures on top of the columns, the wooden rails on the staircase, the glass of the leaded windows, the colored paintings on the dome, and the copper door in the hallway…

Instead of conveying the sentiment for the loss of the good-old-days, the golden age in this book, refers to the period of the end of the 19th and the beginning of the 20th Century when Tianjin, a financial center in the northern part of China, was rich in banks and tremendously prosperous.

This is the real golden age in the history of Tianjin.

Tianjin was first opened up as an international commercial port in 1860. Linking inland China with the rest of the world, Tianjin, with its advantageous geography and convenient land and water communications, attracted merchants from all over the world. Influenced by western styles, the economy of Tianjin soon flourished. The method of trade remittance also transformed from the traditional Chinese method to international standards. The appearance and rapid development of the banking industry was one of the most important facets of Tianjin's development. Foreign-funded banks, in particular, carrying with them the waves of the industrial revolution from capitalist economies, initiated modern financial systems in Tianjin. With the continuous expansion of domestic and international trade, banks were established one after another. These banking institutions from all over the world transformed the city's economy. During this golden age, Tianjin began to play a more and more important role on the economic stage both in China and the World.

Originating from different parts of the world, these banks took on highly individual and unique architectural, decoration and design styles. Exquisite and picturesque the banks greatly enriched the architecture of the city. The physical structures soon became iconic landmarks of this busy commercial port.

Unfortunately due to natural disasters, wars and the Cultural Revolution, few of the banks remain in their original form; moreover. As symbols of western capitalism their importance as city landmarks was also tarnished. Tianjin, one of Chinas great cities, lost a valuable part of her cultural heritage.

In the new millennium, with the opening up of China and with rapidly improving international communications Tianjin is once again forming closer links to the world; For Tianjin to prosper as an economic city, it should reexamine its past, rediscover its heritage, and learn from its cultural, economic and diplomatic history. Even the grand names of these historic banks are precious treasures from the past. They stand as testament to the rich cultural wealth and glorious history of the City of Tianjin. It is thanks to the constant diligence and hard work of local photographers that has made possible this look into Tianjin's golden age.

It is heartening that, while we are researching and rediscovering the historic achieves, new opportunities are knocking at Tianjin's door: the establishment of the China Bohai Bank; the development of Binhai New Area as a special reform area; Tianjin is well positioned to become China's northern economic center. A number of new policies concerning opening-up and financial reform and development have been initiated and tested in Tianjin; foreign-funded banks are being reestablished one after another and are expanding their business in Tianjin…

A new "Golden Age" of prosperous finance and commerce is occurring in Tianjin. Finance, an keystone of the modern economy, is again playing an indispensable role in the renewed growth of Tianjin. Tianjin is fast becoming an attractive financial center for Northern China.

Instead of pondering the memories of the golden age of yesteryear, Tianjin now has the confidence in reaping the riches of tomorrow.

Let's behold her new "Golden Age" together.

后 记
Postscript

《天津老银行》是国内首部有关老银行的画册，我们力图通过这本画册，找出众多建筑遗存中天津在上世纪初作为中国北方金融中心的见证，展现昔日"东方华尔街"的辉煌，从而打开一条天津走向世界的文化通道。

本书共收录了天津现存的老银行遗址35家，几乎囊括了近代天津遗存的所有老银行建筑。其中对18家最具代表性的老银行进行了详细介绍，其余17家采用综合的形式介绍。对于没有旧址的老银行以文字附录的形式记载。总计搜集整理近代天津老银行共有188家，天津老银号、钱庄142家。

本书从选题策划到完稿历时两年半，从17 000余幅图片中精选360余幅编辑成册。照片历时春夏秋冬四季，由外到内，从全景到局部特写，从视角选择到光影运用，无不精心细致，颇具匠心。我们力图使图片不仅有记录历史的价值，还具有一定的审美价值。

原中国人民银行行长、原天津市市长、著名金融家戴相龙先生为本书欣然作序；天津市人民政府外事办公室主任、经济学博士田贵明先生为本书拨冗题跋；著名专家学者周祖奭先生、罗澍伟先生、刘仲直先生、金彭育先生、天津日报武志成先生和作家陈丽伟先生在本书选题策划、文稿编纂、资料提供等方面多有贡献，书中一些珍贵的资料尚属首次公开发表，他们的工作，进一步提高了本书的权威性和专业性，使之兼具了史料价值和收藏价值，在此谨致谢忱。

感谢摄影家高平先生的热诚合作，感谢高大鹏工作室全体同人从灯光造型、装帧设计、编辑等各个方面为此书的出版所付出的辛勤劳动。感谢天津外国语学院常子霞教授、天津财经大学张涌泉教授；感谢天津市钱币收藏学会赵伊先生、天津大学出版社韩振平先生、天津文物局程绍卿先生、英国友人Adam Kerby先生、赵尼诺小姐、王宗发先生、何广宇先生、李旭飞先生、阎丛顺先生、邢振民先生、马彬彬小姐、程桂敏先生、徐英杰先生、徐世勇先生、肖勇先生等在本书翻译、编辑出版及拍摄过程中所给予的帮助。并对以下支持和协作单位一并表示感谢：

天津市科学技术委员会	天津市科学技术协会
天津市和平区邮电局解放路支局	天津国际拍卖有限责任公司
天津市百货商务贸易总公司	天津市和平区繁华地区办公室
中国人民银行天津分行	中国银行天津分行
中国工商银行天津分行	中国建设银行天津分行
中国农业银行天津分行	中国交通银行天津分行
中国农村合作银行天津分行	天津银行和平路分理处
天津金融城开发有限公司	

大鹏文化传播
2007年12月

天津老银行
TIANJIN OLD BANKS

总 顾 问：戴相龙　黄兴国
顾　　问：崔津渡　陈宗胜　郭庆平
主　　办：天津市人民政府外事办公室
编 委 会
主　　任：田贵明
编　　委：王进诚　杜　强　路　红　车德宇
　　　　　　 华耀纲　王　恒　付　钢　曾见泽
　　　　　　 王金龙　李宗唐　齐逢昌　羊子林
　　　　　　 罗澍伟　周祖奭　王树华
统　　筹：赫　维

策　　划：大鹏文化传播
摄　　影：高　平　高大鹏
撰　　稿：周祖奭　刘仲直　金澎育
　　　　　　 武志成　陈丽伟
翻　　译：张涌泉　常子霞　高　雅
英文校译：周祖奭　Adam Kerby
装帧设计：高大鹏工作室

大鹏堂
图说天津文化系列丛书
地　址：天津古文化街文化小城A区205
电　话：27330781　27330167

责任编辑：韩振平
装帧设计：高大鹏工作室

图书在版编目(CIP)数据

天津老银行 / 高大鹏　高平编著. –天津: 天津大学出版社，
2008.3
　ISBN 978-7-5618-2598-3

Ⅰ.天… Ⅱ.①高…②高… Ⅲ.银行-简介-天津市　Ⅳ.
F832.721

中国版本图书馆CIP数据核字(2008)第024296号

出版发行：天津大学出版社
出 版 人：杨　欢
地　　址：天津市卫津路92号天津大学内（邮编：300072）
电　　话：发行部 022-27403647　邮购部 022-27402742
印　　刷：深圳华新彩印制版有限公司
经　　销：全国各地新华书店
开　　本：285mm×400mm
印　　张：39
字　　数：520千
版　　次：2008年2月第1版
印　　次：2008年2月第1次
定　　价：480.00元

版权所有·侵权必究